Springer Theses

Recognizing Outstanding Ph.D. Research

Aims and Scope

The series "Springer Theses" brings together a selection of the very best Ph.D. theses from around the world and across the physical sciences. Nominated and endorsed by two recognized specialists, each published volume has been selected for its scientific excellence and the high impact of its contents for the pertinent field of research. For greater accessibility to non-specialists, the published versions include an extended introduction, as well as a foreword by the student's supervisor explaining the special relevance of the work for the field. As a whole, the series will provide a valuable resource both for newcomers to the research fields described, and for other scientists seeking detailed background information on special questions. Finally, it provides an accredited documentation of the valuable contributions made by today's younger generation of scientists.

Theses may be nominated for publication in this series by heads of department at internationally leading universities or institutes and should fulfill all of the following criteria

- They must be written in good English.
- The topic should fall within the confines of Chemistry, Physics, Earth Sciences, Engineering and related interdisciplinary fields such as Materials, Nanoscience, Chemical Engineering, Complex Systems and Biophysics.
- The work reported in the thesis must represent a significant scientific advance.
- If the thesis includes previously published material, permission to reproduce this must be gained from the respective copyright holder (a maximum 30% of the thesis should be a verbatim reproduction from the author's previous publications).
- They must have been examined and passed during the 12 months prior to nomination.
- Each thesis should include a foreword by the supervisor outlining the significance of its content.
- The theses should have a clearly defined structure including an introduction accessible to new PhD students and scientists not expert in the relevant field.

Indexed by zbMATH.

Olga Movilla Miangolarra

Stochastic Thermodynamic Treatment of Thermal Anisotropy

Doctoral Thesis accepted by the University of California, Irvine, USA

Springer

Olga Movilla Miangolarra
Mechanical and Aerospace Engineering
University of California, Irvine
Irvine, CA, USA

ISSN 2190-5053 ISSN 2190-5061 (electronic)
Springer Theses
ISBN 978-3-031-68065-6 ISBN 978-3-031-68066-3 (eBook)
https://doi.org/10.1007/978-3-031-68066-3

This Springer imprint is published by the registered company Springer Nature Switzerland AG
The registered company address is: Gewerbestrasse 11, 6330 Cham, Switzerland

To my grandmothers

Supervisor's Foreword

The past two decades have witnessed great strides in our ability to effectively model and understand nonequilibrium behavior of thermodynamic systems that are subject to strong thermal fluctuations. These advances have their roots in the molecular basis of thermal properties of matter, postulated more than a century ago, that permits spontaneous violations of the second law of thermodynamics albeit with very low probability. Fluctuation theorems were developed in the 1990s that quantify the strength of the second law, and a new subject was born, finite-time thermodynamics. In this, transitions far from equilibrium are now modeled using Langevin stochastic differential equations and a computational framework has been developed, akin to stochastic control, that is referred to as stochastic energetics or stochastic thermodynamics.

The new theory has been amply validated in the overdamped regime for mesoscopic systems, representing colloidal particles in optical traps and certain biological systems, to accurately quantify entropy production and transference of heat and work between subsystems. Within this setting it has become possible to analyze models of Carnot engines in contact with heat baths of varying temperature, the energetic costs of manipulating nanoparticles, the thermodynamic value of information for control, and to quantify the thermodynamic cost of Landauer's bit erasure taking place in finite time.

In this rapidly developing field, a serendipitous confluence of ideas came along. The mathematics of optimal mass transport, two centuries in the making, provided a natural metric on the space of probability distributions, the so-called Wasserstein metric, that proved to have thermodynamic significance. Thermodynamic states are represented by probability distributions, and the Wasserstein length between them quantifies minimal entropy production to effect a corresponding thermodynamic transition. Important questions of quantifying maximal power and efficiency at maximal power were soon revisited for Carnot-like cycles and fundamental bounds established.

Yet, energy transduction in nature and biology does not typically follow the Carnot cycle that feeds on a periodic driving source, whether thermal or chemical. Instead, mechanisms that include molecular engines, flagellar motors, and the paradigmatic large-scale Stirling engine take advantage of gradients in temperature or chemical potentials to generate motion and produce power. The thesis by Olga Movilla Miangolarra was set out to explore the fundamentals of the underlying mechanism, specifically to explain how thermal engines harvest energy from thermal gradients and to quantify the limits of their ability to produce power.

The thesis begins with an exposition of theory of optimal mass transport. The emphasis is on the induced differential structure of the space of probability distributions, which is then specialized to Gaussian distributions, and provides the mathematical setting for studying cyclic operation of thermodynamic engines. An introduction to stochastic thermodynamics follows with a comprehensive exposition of recent key developments, including original results on power and efficiency of Carnot-like thermal engines, a modification of the second law that takes into account the value of information, and a discussion of detailed fluctuation theorems that concludes with Jarzynski's inequality.

The subsequent three chapters, from the third to the fifth, develop the core theoretical contributions of the thesis. The model of the Brownian gyrator is adopted as an idealization of a thermal engine that couples two heat baths of different temperature via the system dynamics. The anisotropy in the thermal excitation produces heat flux and a torque that generates power for the gyrating engine to transfer to an external subsystem. The amount of energy that is being harvested during a cycle is expressed as an area integral over a domain in the Wasserstein space that is enclosed by the closed path traversed by the thermodynamic system. The dissipated energy during the cycle is quantified by the Wasserstein length of the path. Thereby, power optimizing cycles amount to solutions of an isoperimetric problem, to maximize area for a specified perimeter of a cycle or, equivalently, minimize the perimeter given the area of the enclosed domain. The physical parameter that differentiates optimal cycles is their period—when the specified period is too short, dissipation may exceed the work drawn and the gyrating engine produces no net power. Further, entropy production due to the seepage of heat between the heat baths through the system dynamics is studied for a variety of operating conditions.

An embodiment of a model thermal engine is developed in Chap. 6. The engine is in contact with heat baths of different temperature and feeds on the thermal gradient. It consists of a system of time-varying capacitors with voltages and currents that are stochastically driven by Johnson-Nyquist thermal noise, which is in turn supplied by thermal resistors maintained at these different temperatures. Dielectric materials in the capacitors act as the pistons of the engine. They are mechanically coupled to an inertial wheel that transfers energy to an external mechanical load. The design constitutes a real-life embodiment of a Brownian gyrator turned into a thermal engine. Numerical experimentation and theoretical analysis establish the stability of its limit cycle operation. The chapter concludes with an inertialess concept, where parallel coupling of multiple engines with a phase difference in their cyclic operation removes the need for inertia to maintain stable operation.

The thesis by Olga Movilla Miangolarra provides a proof-of-concept for power producing engines that feed on thermal gradients and develops the theory to quantify the limits to their performance. The elegance and sharpness of the results that, in particular, quantify maximal power via isoperimetric inequalities within the edifice of the Wasserstein geometry, represent a major advance in our understanding of energy transduction by thermal engines that operate far from equilibrium.

Irvine, CA, USA Tryphon T. Georgiou

Acknowledgments

I would like to express my deepest gratitude to Prof. Tryphon T. Georgiou for his invaluable guidance, in and out of science, and for sharing with his students the most valuable thing we own—our time. I want to thank Prof. Amir Taghvaei; I can wholeheartedly say that without his help, none of this would have been possible. To Prof. Yongxin Chen for countless meetings, words of wisdom, and especially for pointing me to the topic of anisotropic fluctuations. I am grateful to my labmates, especially Dr. Rui Fu and Jordi Ventura, for their work, their help, and their companionship.

To my family, Ama, Aita, Ander, for their constant support. To my friends in California, and especially Zeke, for without them, this thesis would have probably still been possible, but with all certainty a lot more boring. To my friends back home, for always being there for me.

A special thanks is also in order to Pete Balsells and to "La Caixa" Foundation, who have ultimately enabled the writing of these words. Additional funding from NSF under grant 1807664, the AFOSR under FA9550-23-1-0096, and ARO under W911NF-22-1-0292 is gratefully acknowledged.

Parts of This Thesis Have Been Published in the Following Journal Articles

O. Movilla Miangolarra, A. Taghvaei, T. T. Georgiou, "Minimal entropy production in the presence of anisotropic fluctuations"
IEEE Transactions on Automatic Control, 2025

J. Ventura Siches, O. Movilla Miangolarra, T. T. Georgiou, "Refined bounds on energy harvesting from anisotropic fluctuations"
Physical Review E, 2024

R. Fu, O. Movilla Miangolarra, A. Taghvaei, Y. Chen, T. T. Georgiou, "Stochastic thermodynamic engines under time-varying temperature profile"
Automatica, 2024

O. Movilla Miangolarra, A. Taghvaei, T. T. Georgiou, "A matching principle for power transfer in Stochastic Thermodynamics"
IEEE Control Systems Letters, 2023

O. Movilla Miangolarra, A. Taghvaei and T. T. Georgiou, "Minimal entropy production in anisotropic temperature fields"
IEEE American Control Conference (ACC), 2023

J. Ventura Siches, O. Movilla Miangolarra, A. Taghvaei, Y. Chen, T. T. Georgiou, "Inertialess Gyrating Engines"
PNAS Nexus, 2022

O. Movilla Miangolarra, A. Taghvaei, Y. Chen, T.T. Georgiou, "Geometry of finite-time thermodynamic cycles with anisotropic thermal fluctuations"
IEEE Control Systems Letters, 2022

O. Movilla Miangolarra, A. Taghvaei, Y. Chen, T. T. Georgiou, "Thermodynamic engine powered by anisotropic fluctuations"
Physical Review Research, 2022

O. Movilla Miangolarra, A. Taghvaei, R. Fu, Y. Chen, T. T. Georgiou, "Energy harvesting from anisotropic fluctuations"
Physical Review E, 2021

O. Movilla Miangolarra, R. Fu, A. Taghvaei, Y. Chen, T. T. Georgiou, "Underdamped stochastic thermodynamic engines in contact with a heat bath with arbitrary temperature profile"

Physical Review E, 2021

A. Taghvaei, O. Movilla Miangolarra, R. Fu, Y. Chen, T. T. Georgiou, "On the relation between information and power in stochastic thermodynamic engines"

IEEE Control Systems Letters, 2021

R. Fu, O. Movilla Miangolarra, A. Taghvaei, Y. Chen and T. T. Georgiou, "Harvesting energy from a periodic heat bath"

IEEE Conference on Decision and Control (CDC), 2020

Preface

Ars longa, vita brevis
– Hippocrates.

The second law of thermodynamics states that the entropy of an isolated system can never decrease in time. Entropy can be thought of as a measure of chaos, uniformity, disorder. So the second law is the reason why your room is not tidy. It is the reason why we feel time passing, why we are able to *experience* being alive. But, at the same time, the second law should have been an obstacle to the existence of life—life is characterized by the ability to create structure, order. We are complex systems, our bodies, our molecules, built of atoms precisely aligned, are very far from being just a disordered lump of matter.

Yet here we are, alive, in all of our structure and complexity, reading these words. This is only possible because we are not isolated systems. We feed from the entropy of our environments, from the food we eat, the air we breathe, and the water we drink [1]. We use these to generate energy that can fight back entropic forces, maintaining the integrity and homeostasis[1] of our bodies [2]. Then, we discard the waste products back into the environment, thus increasing its entropy. Hence, entropy is being produced *globally*, while it is *locally* decreased through the usage of energy.

This is only possible thanks to mechanisms of energy generation at the expense of the environment's entropy, which are ubiquitous in living systems. Common examples of such mechanisms are the $F0/F1$ ATP synthase and the bacterial flagellar motor [3]. In most cases, it is a gradient of ion (or other chemical) concentration or temperature, what provides the fuel that feeds these dynamical processes; the enabling mechanism responsible for transduction of energy is typically the presence of a non-equilibrium steady state (NESS) [4, 5]. Thus, the anisotropy generated by these gradients and the non-equilibrium states that it induces are the hallmark of life, and living systems by necessity operate away from equilibrium.

[1] The state of steady internal conditions in living systems.

Hence, understanding non-equilibrium thermodynamics at the microscopic level for systems in such anisotropic environments is of *vital* importance in our understanding of biological processes, and ultimately, in our understanding of life. This thesis is devoted to that subject, in particular, to the exploration of how thermodynamic systems subject to anisotropic temperatures can optimize their operation in order to make the most of the available resources.

The outline of this thesis is as follows. First, we introduce the theory of Optimal Mass Transport (OMT), which will constitute the backbone of this manuscript, giving us the necessary mathematical tools to tackle our problems. Then, in Chap. 2, we will introduce the basic elements of Stochastic Thermodynamics, a theory that represents a modern version of classical thermodynamics that is able to describe thermodynamic systems at the microscale, and which will constitute the framework of this thesis. Having laid down the necessary background, we will dive into a central problem in thermodynamics, that of maximizing work extraction, which will be studied in Chap. 4. Therein, the anisotropy in temperature will provide the fuel enabling energy harvesting. Chapter 5 will be devoted to the problem of minimizing entropy production, where anisotropy will play a leading role, intrinsically altering the nature of entropy generation. Finally, in Chap. 6, we will apply the gained knowledge to give an example of an autonomous engine that is able to stably extract energy from the anisotropic environment, the sought insights aiming at nano-engines and biological motors.

Irvine, CA, USA Olga Movilla Miangolarra

References

1. Schrödinger, E.: The physical aspect of the living cell. What is Life 70–71 (1944)
2. Movilla Miangolarra, A.: Intracellular Spatial Organisation: The Effects of Non-ideal Solutions. Theses, Université Paris sciences et lettres, 2021
3. Oster, G., Wang, H.: Rotary protein motors. Trends Cell Biol. **13**(3), 114–121 (2003)
4. Battle, C., Broedersz, C.P., Fakhri, N., Geyer, V.F., Howard, J., Schmidt, C.F., MacKintosh, F.C.: Broken detailed balance at mesoscopic scales in active biological systems. Science (2016)
5. Gnesotto, F.S., Mura, F., Gladrow, J., Broedersz, C.P.: Broken detailed balance and non-equilibrium dynamics in living systems: a review. Rep. Progress Phys. (2018)

Contents

Chapter 1
Introduction to Optimal Mass Transport

1.1 The Monge Problem

Optimal Mass Transport (OMT) was born from a question posed by Gaspard Monge in 1781: how to transport a pile of sand from one place to another, while minimizing some transportation cost [1]. This pile of sand can be modeled through a probability measure by normalizing its mass to one. So the task at hand can be formulated as that of optimally transporting a pile of sand from an initial distribution μ to a final distribution ν.

Let us denote by $P(X)$ the space of probability measures on X. For simplicity, throughout this manuscript, we restrict ourselves to the space of probability measures with finite second-order moments on $\mathbb{R}^n$, i.e,

$$P_2(\mathbb{R}^n) := \left\{ \lambda \in P(\mathbb{R}^n) \; : \; \int_{\mathbb{R}^n} xx' d\lambda(x) < +\infty \right\},$$

where $'$ denotes transpose. In addition, these probability measures are assumed absolutely continuous with respect to the Lebesgue measure, this is, are assumed to assign zero measure to sets with zero Lebesgue measure. This last property implies that probability measures can be identified with probability densities.[1] We would like to note that these restrictions are not always necessary and the reader is referred to [2] for the nitty-gritty details.

[1] This is done through a concept referred to as the Radon-Nikodym derivative. Specifically, if μ is absolutely continuous with respect to the Lebesgue measure, then there exists a density ρ such that $d\mu(x) = \rho(x)dx$, and ρ is precisely the Radon-Nikodym derivative of μ with respect to the Lebesgue measure.

O. Movilla Miangolarra, *Stochastic Thermodynamic Treatment of Thermal Anisotropy*, Springer Theses, https://doi.org/10.1007/978-3-031-68066-3_1

Fig. 1.1 The Monge problem

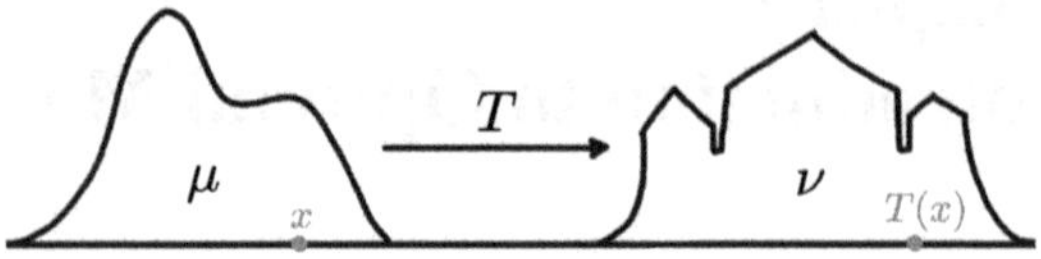

Monge wanted to transport the sand through a transport map $T : \mathbb{R}^n \to \mathbb{R}^n$, that takes the grains of sand at position $x \in \mathbb{R}^n$ to position $y = T(x) \in \mathbb{R}^n$ (see Fig. 1.1). This transport map must be such that when it "acts" on μ it gives us ν back. This fact can be precisely captured by the push-forward of μ with respect to T, this is,

$$T_{\#}\mu = \nu,$$

where the push-forward is defined as $(T_{\#}\mu)(A) := \mu(T^{-1}(A))$, for all measurable sets $A \subset \mathbb{R}^n$.

In this way, we can formulate the Monge problem as follows:

Monge Problem
Let $\mu \in P_2(\mathbb{R}^n)$ and $\nu \in P_2(\mathbb{R}^n)$ be absolutely continuous measures with respect to the Lebesgue measure, and $d : \mathbb{R}^n \times \mathbb{R}^n \to [0, \infty)$ be a specified transportation cost. Determine

$$\inf_T \left\{ \int_{\mathbb{R}^n} d(x, T(x)) d\mu(x) \mid T_{\#}\mu = \nu \right\}.$$

While the original question by Monge aimed at taking the Euclidean distance $d(x, y) = \|x - y\|$ as the transportation cost, more general types of candidate costs have since been considered. In this thesis, we will be mostly concerned with the square of the Euclidean distance $d(x, y) = \|x - y\|^2 = (x - y)'(x - y)$ as the transportation cost. This choice turns out to have nice physical consequences that we will see later on. In addition, it is conveniently strictly convex and defines the square of a metric on the space of measures, which leads to a rather rich structure [2].

The Monge problem drew a lot of attention at the time. Indeed, almost a century later, the Academy of Paris offered a prize for its solution [2]. However, even then, it was met with limited success. This was mostly due to the nonlinear nature of the push-forward constraint. It wasn't until another century later that Leonid Kantorovich was able to pose this problem in a slightly weaker way, subject to only linear constraints. This opened the gate to a myriad of new results and allowed the field to flourish.

1.2 The Kantorovich Problem

Kantorovich relaxed the assumption of having all particles at position x being transported to exactly the same position y, by allowing what we call splitting of mass. That is, particles starting at the same location can be assigned different destinations in general. This relaxation of the problem circumvents technical difficulties linked to non-absolutely continuous measures that frustrated early attempts on the Monge problem. Thus, Kantorovich's idea was to skip looking for an optimal transport map T, that rigidly transports mass from a given location x to the same destination y, and instead consider probability measures π on the product space $\mathbb{R}^n \times \mathbb{R}^n$ that quantify the amount of mass that travels from position x to possibly more than one location y (see Fig. 1.2, left). Such a measure is referred to as coupling of the two marginals μ and ν. Then, one has to look in the space of couplings, denoted by

$$\Pi(\mu, \nu) := \left\{ \pi \in P(\mathbb{R}^n \times \mathbb{R}^n) : \int_{\mathbb{R}^n} \pi \, dy = \mu, \int_{\mathbb{R}^n} \pi \, dx = \nu \right\},$$

for a minimizer of the transportation cost, as expressed below, specialized to the cost of interest $\|x - y\|^2$.

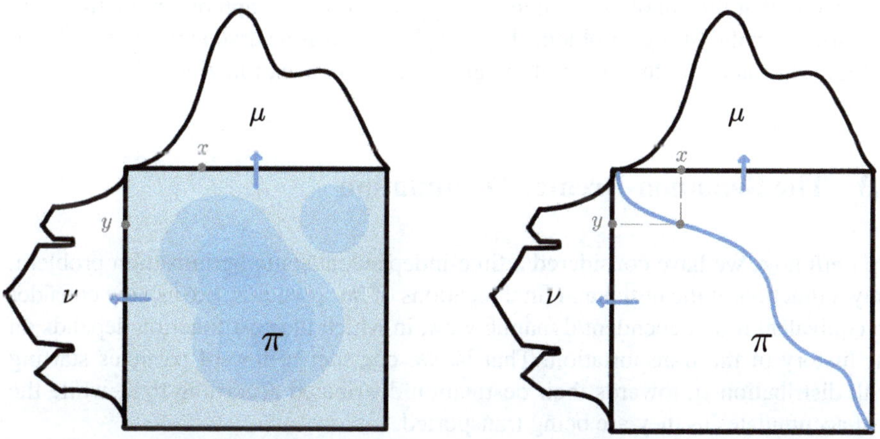

Fig. 1.2 The Kantorovich problem. Left: The coupling $\pi \in \Pi(\mu, \nu)$ is portrayed as a probability distribution on the product space $\mathbb{R}^n \times \mathbb{R}^n$ with μ and ν as marginals. Right: The optimal coupling has its support on a curve which relates x to y in a one-to-one fashion. This coupling can be given by a transport map T, and thus the solutions to the Monge and Kantorovich problems coincide

Kantorovich Problem
Let $\mu \in P_2(\mathbb{R}^n)$ and $\nu \in P_2(\mathbb{R}^n)$ be absolutely continuous measures with respect to the Lebesgue measure, determine

$$\inf_{\pi \in \Pi(\mu,\nu)} \int_{\mathbb{R}^n \times \mathbb{R}^n} \|x - y\|^2 d\pi(x, y).$$

The great new element of Kantorovich's formulation is that in great generality, and for all kinds of different costs, the constrains on π are linear, and so is the integral cost to be optimized.

By replacing the Kantorovich problem with its dual problem, one can show (after some work, see [2, Chapters 1 and 2]) that the Kantorovich problem has a unique solution (for non-necessarily absolutely continuous marginals, and under very weak assumptions). The minimal value, referred to as the 2-Wasserstein distance between measures μ and ν, and denoted by $\mathrm{W}_2(\mu, \nu)$, defines the square of a metric[2] on the space of probability measures with finite second-order moments, $P_2(\mathbb{R}^n)$.

It turns out that when μ, ν are absolutely continuous as assumed herein, the solution to the Kantorovich problem coincides with the solution to the Monge problem in that the coupling obtained is supported on the graph of a map, the same map sought in the Monge problem. This graph, shown as a curve in the right subplot of Fig. 1.2, is actually the graph of the gradient of a convex function [2].

1.3 The Benamou-Brenier Formulation

Up until now, we have considered a time-independent/static optimization problem, only a function of the initial and final locations of our particles. Let us now consider an equivalent time-dependent/dynamic view, in which the cost function depends on the history of the transportation. That is, we consider a flow of particles starting with distribution μ, towards their destination distributed according to ν, while the cost accumulates as they are being transported.

We first consider a Lagrangian point of view in which each initial position x is associated with a trajectory $\{X_t(x)\}_{0 \le t \le t_f}$, where t_f is the final time by which we wish to complete the transportation, and a velocity $v(t, X_t(x))$, related through

$$\partial_t X_t(x) = v(t, X_t(x)), \quad X_0(x) = x.$$

[2] Being a metric, $\mathrm{W}_2(\cdot, \cdot)$ gives a positive value when its two arguments differ and zero otherwise, it is symmetric in that it is unaffected by their order, and obeys the triangle inequality.

In order to find a time-dependent problem whose solution coincides with the Monge problem solution, let us first solve the following optimization problem

$$\inf_{X_t}\left\{t_f\int_0^{t_f}\|\partial_t X_t\|^2 dt \mid X_0(x)=x,\ X_{t_f}(x)=y\right\}.$$

One can write the first-order optimality condition to obtain that

$$\partial_t^2 X_t = 0 \implies X_t(x) = x + \frac{t}{t_f}(y-x),$$

this is, optimal trajectories have constant velocity, and

$$\|x-y\|^2 = \inf_{X_t}\left\{t_f\int_0^{t_f}\|\partial_t X_t\|^2 dt \mid X_0(x)=x,\ X_{t_f}(x)=y\right\}.$$

Because the solution to this problem is the distance function that we are trying to minimize, it is not surprising that the Monge problem can be written in Lagrangian formulation as

$$\inf_{X_t}\left\{t_f\int_0^{t_f}\int_{\mathbb{R}^n}\|\partial_t X_t(x)\|^2 d\mu(x)dt \mid X_0=\text{Id},\ (X_{t_f})_\#\mu=\nu\right\},$$

where Id is the identity map. Thus, when we look at the Monge problem as a time-dependent problem, we obtain that the optimal trajectories are straight lines, parametrized with constant velocity in the sense of this Lagrangian viewpoint.

As is the case of any fluid dynamical description, here too we can switch from this Lagrangian formulation to an Eulerian one. Indeed, let $\rho(t,x)$ describe the amount of particles (or equivalently, probability mass) at position x (seen now as a fixed point on $\mathbb{R}^n$) and time t. Let it be given by $\rho(t,\cdot) = (X_t)_\#\mu$, where $X_t(x) = x + tv(t,x)$, with $v(t,x)$ Lipschitz. Note that here, $v(t,x)$ represents the velocity of any particle located at position x at time t (as opposed to $v(t, X_t(x))$ which represents the velocity of the particle that *started* at x after time t). Then, it turns out that ρ is the unique (weak) solution to

$$\partial_t\rho = -\nabla\cdot(\rho v), \qquad \rho(0,x)dx = d\mu(x),$$

where $\nabla\cdot$ denotes the divergence with respect to the x variable. We will call this equation the continuity equation, since it enforces mass conservation, i.e. $\int_{\mathbb{R}^n}\rho(t,x)dx = 1$ for all t.

With this in mind one has

$$\int_0^{t_f}\int_{\mathbb{R}^n}\|v(t,x)\|^2\rho(t,x)dxdt = \int_0^{t_f}\int_{\mathbb{R}^n}\|v(t,X_t(x))\|^2\rho(0,x)dxdt$$
$$= \int_0^{t_f}\int_{\mathbb{R}^n}\|\partial_t X_t\|^2 d\mu dt,$$

which leads us to our final formulation of the OMT problem, the Benamou-Brenier formulation [3]:

Benamou-Brenier Formulation

Let $\mu \in P_2(\mathbb{R}^n)$ and $\nu \in P_2(\mathbb{R}^n)$ be absolutely continuous measures with respect to the Lebesgue measure, then the Monge (and Kantorovich) problem is equivalent to

$$\mathrm{W}_2(\mu,\nu)^2 = \min_{(\rho,v)}\left\{t_f\int_0^{t_f}\int_{\mathbb{R}^n}\|v(t,x)\|^2\rho(t,x)dxdt \mid \partial_t\rho = -\nabla\cdot(\rho v)\right\}, \tag{1.1}$$

with $\rho(0,x)dx = d\mu(x)$ and $\rho(t_f,x)dx = d\nu(x)$.

In this way, the optimal mass transport problem can be thought of as a minimization of an action integral; the integral over time of the kinetic energy of the particles that flow according to the continuity equation from distribution μ to ν.

Let us assume that v_0 minimizes problem (1.1). Then, for any $\epsilon \neq 0$, $\widetilde{v} = v_0 + \epsilon\chi$ is an admissible velocity field (meaning it also satisfies the continuity equation), provided that χ is a divergence-free vector field with respect to ρ, i.e. $\nabla\cdot(\rho\chi) = 0$. Consider,

$$\int_0^{t_f}\int_{\mathbb{R}^n}\|v_0+\epsilon\chi\|^2\rho dxdt = \int_0^{t_f}\int_{\mathbb{R}^n}\|v_0\|^2\rho dxdt + 2\epsilon\int_0^{t_f}\int_{\mathbb{R}^n}\langle v_0,\chi\rangle\rho dxdt$$
$$+\epsilon^2\int_0^{t_f}\int_{\mathbb{R}^n}\|\chi\|^2\rho dxdt,$$

where $\langle u, v\rangle = u'v$ denotes the standard Euclidean inner product with $'$ denoting transpose. Note that the last term is always positive, while the second to last need not be. Since v_0 is minimizing,

$$\int_0^{t_f}\int_{\mathbb{R}^n}\|v_0\|^2\rho dxdt \leq \int_0^{t_f}\int_{\mathbb{R}^n}\|v_0+\epsilon\chi\|^2\rho dxdt \quad \text{for all } \epsilon \in \mathbb{R}, \tag{1.2}$$

which implies that

$$\int_{\mathbb{R}^n} \langle v_0, \chi \rangle \rho dx = 0 \quad \text{for any divergence-free } \chi,$$

at all times. Therefore, optimal velocities must be orthogonal (with respect to ρ) to divergence-free vector fields. This fact will be crucial to define a differential structure in our space of probability distributions, as we will see in what follows.

1.4 A Differential Point of View

We can use Cauchy-Schwarz inequality to lower bound the transportation cost being minimized in (1.1)

$$t_f \int_0^{t_f} \int_{\mathbb{R}^n} \|v(t,x)\|^2 \rho(t,x) dx dt \geq \left(\int_0^{t_f} \sqrt{\int_{\mathbb{R}^n} \|v(t,x)\|^2 \rho dx}\, dt \right)^2 =: \ell^2, \tag{1.3}$$

where equality is achieved for constant velocities v along a trajectory, as we have previously seen. The expression inside the parenthesis can be thought of as a length, ℓ, in the space of probability distributions. Therefore, the problem in (1.1) is nothing but that of finding the path of minimal length that joints two points, μ and ν, in the space of probability distributions. Its solution gives the geodesics on that space.

This idea can be made formal by defining an inner product on the tangent space of the space of probability distributions. In order to do so we additionally consider the following mild assumptions on thermodynamic states and corresponding vector-fields:

Assumption A1 For all probability densities $\rho(x)$ the following hold:

$$\text{(i)} \quad \|\nabla^2 \log \rho\|_\infty = \sup_{x \in \mathbb{R}^n} |\nabla^2 \log \rho(x)| < \infty$$

$$\text{(ii)} \quad \|\nabla \log \rho(x)\| \to \infty \text{ as } \|x\| \to \infty.$$

Assumption A2 Vector-fields $v(t,x)$ are Lipschitz, i.e. there exists a $\kappa > 0$ such that for all $(t_1, x_1), (t_2, x_2) \in \mathbb{R} \times \mathbb{R}^n$ we have

$$\|v(t_2, x_2) - v(t_1, x_1)\| \leq \kappa(\|x_2 - x_1\| + |t_2 - t_1|).$$

Assumption A1 ensures that ρ satisfies the Poincaré inequality[3] [4], i.e. that there exists a positive constant $C > 0$ such that

[3] The Poincaré inequality essentially expresses that 0 is an isolated eigenvalue of a corresponding Laplace operator.

$$\int |h - \bar{h}|^2 \rho dx \leq C \int \|\nabla h\|^2 \rho dx \text{ for all } h \in H^1_\rho,$$

where $\bar{h} = \int h\rho dx$ and H^1_ρ is the Sobolev space of functions where the function and its derivative, defined in the weak sense, are square integrable with respect to ρ, namely, that both $\int h^2 \rho dx$, and $\int \|\nabla h\|^2 \rho dx$ are bounded. The Poincaré inequality is of importance as it provides a sufficient condition for existence and uniqueness of the solution $\phi \in H^1_\rho$ to the (weighted) Poisson equation

$$\mathcal{L}_\rho \phi = \rho h,$$

where $\mathcal{L}_\rho(\cdot) := \nabla \cdot (\rho \nabla(\cdot))$ is the (weighted) Laplacian and $\int h^2 \rho dx < \infty$ [4]. The relevance of the existence and uniqueness of solutions to the Poisson equation will become apparent shortly.

Consider now the following inner product between vector fields on $\mathbb{R}^n$

$$\langle v_1, v_2 \rangle_\rho = \int \langle v_1(x), v_2(x) \rangle \rho(x)\, dx, \tag{1.4}$$

with induced norm $\|v\|_\rho := \sqrt{\langle v, v \rangle_\rho}$. The inner product defines an orthogonal decomposition

$$v = \Pi_\rho\, v + \chi,$$

where Π_ρ is the projection operator given by

$$\Pi_\rho v := \arg\min_{v_0} \{\|v_0\|_\rho \mid v_0 = v - \chi \text{ and } \nabla \cdot (\rho \chi) = 0\}.$$

The projection is unique and belongs to the closure of the space of vector fields of gradient form with respect to the $\|\cdot\|_\rho$ topology [5, Lemma 8.4.2]. Under assumptions A1 and A2, the projection is exactly of gradient form

$$\Pi_\rho v = \nabla \phi, \text{ where } \phi \in H^1_\rho \text{ solves } \mathcal{L}_\rho \phi = \nabla \cdot (\rho v),$$

the (weighted) Poisson equation. The decomposition

$$v = \nabla \phi + \chi,$$

into the gradient and divergence-free parts, is an instance of the known Helmholtz-Hodge decomposition [6, 7].

This decomposition can be used to construct the tangent space at any $\rho \in P_2(\mathbb{R}^n)$. In particular, for an admissible rate of change $\dot{\rho} = -\nabla \cdot (\rho v)$, induced by the vector field v, there is a corresponding gradient vector field $\nabla \phi = \Pi_\rho\, v$, which constitutes the tangent vector. The correspondence between $\dot{\rho}$ and $\nabla\phi$ is used to equip $P_2(\mathbb{R}^n)$ with the Riemannian metric

$$\langle \dot{\rho}_1, \dot{\rho}_2 \rangle_{g_W} := \langle \nabla\phi_1, \nabla\phi_2 \rangle_{\rho},$$

which allows computing lengths of paths between densities. The smallest distance (geodesic) between any two given densities μ and ν is precisely the 2-Wasserstein metric, which can now be written as

$$\mathrm{W}_2(\mu, \nu)^2 = t_f \min_{\rho} \int_0^{t_f} \|\dot{\rho}\|_{g_W}^2 dt \tag{1.5}$$

subject to $\rho(0, x)dx = d\mu(x)$, $\rho(t_f, x)dx = d\nu(x)$.

Let us now recapitulate what we have learnt. We have seen how the OMT problem defines the so-called 2-Wasserstein metric. This induces a Riemannian metric g_W, on the tangent space of probability distributions on $\mathbb{R}^n$ with finite second-order moments ($P_2(\mathbb{R}^n)$), making $P_2(\mathbb{R}^n)$ into a geodesic space. Geodesics correspond to (optimal) flows between endpoint distributions and $\mathrm{W}_2(\mu, \nu)$ becomes the geodesic length. The optimal velocity fields turn out to be curl-free and, hence, gradients $v = \nabla\phi$. These gradients can be identified with tangent vectors of $P_2(\mathbb{R}^n)$.

A word of caution is in order. Herein we have prioritized intuition over rigor. None of this discussion should be taken as an actual proof. For the real deal, the reader is suggested to start with [8] and to look at [2] and [5] for completeness.

Before we switch to the subject of this thesis -Stochastic Thermodynamics-, let us focus on a particular OMT problem that turns out to be explicitly solvable: the Gaussian problem.

1.5 The Gaussian Case: An Explicitly Solvable Example

Let $d\mu(x) = \rho_x(x)dx$ and $d\nu(y) = \rho_y(y)dy$ be two n-dimensional Gaussian probability measures with mean m_x and $m_y \in \mathbb{R}^n$, and covariance matrix Σ_x and $\Sigma_y \in \mathbb{R}^{n\times n}$, respectively. This is, the probability densities ρ_x and ρ_y are given by

$$\rho_i(z) = \frac{1}{(2\pi)^{n/2}\det(\Sigma_i)^{1/2}} e^{-\frac{1}{2}(z-m_i)'\Sigma_i^{-1}(z-m_i)}, \quad \text{for } i \in \{x, y\},$$

where $\det(\Sigma)$ denotes the determinant of Σ. The Kantorovich problem can be thought of as the minimization of the covariance of the coupling π,

$$\min_{\pi\in\Pi(\mu,\nu)} \int_{\mathbb{R}^n\times\mathbb{R}^n} \|x - y\|^2 d\pi(x, y) = \min_{\pi\in\Pi(\mu,\nu)} \mathrm{Tr}\left[\Sigma_x + \Sigma_y - 2\Sigma_{xy}\right],$$

where Tr[·] denotes the trace (remember that any scalar is equal to its trace). Here, Σ_x and Σ_y are fixed by our marginal constraint, i.e. the fact that $\pi \in \Pi(\mu, \nu)$. Therefore, the Kantorovich problem reduces to a simple maximization of $\mathrm{Tr}[\Sigma_{xy}]$ such that the joint covariance matrix of π is positive definite, this is,

$$\begin{aligned}
\max\{\mathrm{Tr}[\Sigma_{xy}] \mid \begin{bmatrix} \Sigma_x & \Sigma_{xy} \\ \Sigma'_{xy} & \Sigma_y \end{bmatrix} \geq 0\} &\Leftrightarrow \max\{\mathrm{Tr}[\Sigma_{xy}] \mid \| \underbrace{\Sigma_x^{-1/2} \Sigma_{xy} \Sigma_y^{-1/2}}_{M} \| \leq 1\} \\
&\Leftrightarrow \max\{\mathrm{Tr}[\Sigma_y^{1/2} \Sigma_x^{1/2} M] \mid \|M\| \leq 1\} \\
&\Leftrightarrow \max\{\mathrm{Tr}[u \sigma v' M] \mid \|M\| \leq 1\} \\
&\Leftrightarrow \max\{\mathrm{Tr}(\sigma \underbrace{v' M u}_{Q}) \mid \|Q\| \leq 1\} \\
&\Leftrightarrow \mathrm{Tr}(\sigma) \text{ for } Q = \mathrm{Id},
\end{aligned}$$

where we have used the fact that $\Sigma_y^{1/2} \Sigma_x^{1/2}$ has a singular value decomposition, $\Sigma_y^{1/2} \Sigma_x^{1/2} = u \sigma v'$. It follows from $Q = \mathrm{Id}$, where Id denotes the identity matrix, that $M = v u'$. One can use the fact that any matrix K with a singular value decomposition of $v_K \sigma_K u'_K$ obeys

$$K(K'K)^{-1/2} = v_K u'_K,$$

to write the optimal Σ_{xy} as

$$\begin{aligned}
\Sigma_{xy} = \Sigma_x^{1/2} M \Sigma_y^{1/2} &= \Sigma_x^{1/2} N (N'N)^{-1/2} \Sigma_y^{1/2} \\
&= \Sigma_x^{1/2} \Sigma_x^{1/2} \Sigma_y^{1/2} (\Sigma_y^{1/2} \Sigma_x \Sigma_y^{1/2})^{-1/2} \Sigma_y^{1/2} \\
&= \Sigma_x \Sigma_y^{1/2} (\Sigma_y^{1/2} \Sigma_x \Sigma_y^{1/2})^{-1/2} \Sigma_y^{1/2},
\end{aligned}$$

where we have used $N = \Sigma_x^{1/2} \Sigma_y^{1/2}$.

Note that the inequality constraint being saturated implies the joint covariance matrix is singular (has zero determinant), which in turn means it can be described through a Monge map. Specifically, the joint covariance matrix can be written as

$$\Sigma = \begin{bmatrix} \mathrm{Id} \\ A \end{bmatrix} \Sigma_x \begin{bmatrix} \mathrm{Id} & A \end{bmatrix},$$

for some matrix $A \in \mathbb{R}^{n \times n}$. We now check that this can be obtained through a linear transport map that gives us back the correct optimal joint coupling Σ. The uniqueness of solutions to the Kantorovich problem will then imply that this is the unique optimizing Monge map.

Indeed, let us try the following transport map:

$$T : x \to y = m_y + A(x - m_x).$$

If x is a Gaussian random variable, then so will y. Its mean will be given by m_y and its covariance by

$$A\Sigma_x A'.$$

Let $A = \Sigma'_{xy}\Sigma_x^{-1} = \Sigma_y^{1/2}(\Sigma_y^{1/2}\Sigma_x\Sigma_y^{1/2})^{-1/2}\Sigma_y^{1/2}$, then

$$A\Sigma_x A' = \Sigma_y^{1/2}(\Sigma_y^{1/2}\Sigma_x\Sigma_y^{1/2})^{-1/2}\Sigma_y^{1/2}\Sigma_x\Sigma_y^{1/2}(\Sigma_y^{1/2}\Sigma_x\Sigma_y^{1/2})^{-1/2}\Sigma_y^{1/2} = \Sigma_y.$$

Thus, the y defined through this transport map T satisfies the required conditions.
Linear interpolation will then give optimal trajectories, i.e.

$$X_t(x) = x + \frac{t}{t_f}(A(x - m_x) + m_y - x) = \frac{t}{t_f}(m_y - Am_x) + \left(\left(1 - \frac{t}{t_f}\right)\mathrm{Id} + \frac{t}{t_f}A\right)x.$$

Taking the expected value of this expression we obtain the mean

$$m_t = m_x + (m_y - m_x)\frac{t}{t_f}.$$

On the other hand, the covariance can be also computed to be

$$\begin{aligned}\Sigma_t &= \left(\left(1 - \frac{t}{t_f}\right)\mathrm{Id} + \frac{t}{t_f}A\right)\Sigma_x\left(\left(1 - \frac{t}{t_f}\right)\mathrm{Id} + \frac{t}{t_f}A\right)' \\ &= \Sigma_x^{-1/2}\left(\left(1 - \frac{t}{t_f}\right)\Sigma_x + \frac{t}{t_f}(\Sigma_x^{1/2}\Sigma_y\Sigma_x^{1/2})^{1/2}\right)^2\Sigma_x^{-1/2}.\end{aligned} \tag{1.6}$$

Thus, we have found an optimal coupling π and transport map T that minimize our transportation cost, giving a distance between our two Gaussians of

$$\mathrm{W}_2(\mu, \nu)^2 = \mathrm{Tr}[\Sigma_x + \Sigma_y - 2(\Sigma_x^{1/2}\Sigma_y\Sigma_x^{1/2})^{1/2}]. \tag{1.7}$$

We will shortly see how this distance is related to thermodynamics and, specifically, the second law. Before we embark on that endeavor, let us introduce some key concepts regarding Stochastic Thermodynamics.

References

1. Monge, G.: Mémoire sur la théorie des déblais et des remblais. Mem. Math. Phys. Acad. R. Sci. 666–704 (1781)
2. Villani, C.: Topics in Optimal Transportation. American Mathematical Society, Providence (2003)
3. Benamou, J.-D., Brenier, Y.: A computational fluid mechanics solution to the monge-kantorovich mass transfer problem. Numer. Math. **84**(3), 375–393 (2000)
4. Laugesen, R.S., Mehta, P.G., Meyn, S.P., Raginsky, M.: Poisson's equation in nonlinear filtering. SIAM J. Control Optim. (2015)
5. Ambrosio, L., Gigli, N., Savaré, G.: Gradient Flows: In Metric Spaces and in the Space of Probability Measures. Springer Science & Business Media, Berlin (2005)
6. Morse, P.M., Feshbach, H.: Methods of theoretical physics. Int. Series Pure Appl. Phys. (1953)
7. Chorin, A.J., Marsden, J.E., Marsden, J.E.: A Mathematical Introduction to Fluid Mechanics, vol. 3. Springer, Berlin (1990)
8. Santambrogio, F.: Optimal transport for applied mathematicians. Birkäuser, NY **55**(58–63), 94 (2015)

Chapter 2
Introduction to Stochastic Thermodynamics

Thermodynamics is the branch of physics concerned with energy exchange. It was born from Carnot's quest to quantify the maximal efficiency of heat engines, which was marvelously accomplished in his celebrated work [1, 2], and led to the discovery of the concepts of entropy, absolute temperature, and irreversibility. Specifically, Carnot introduced the second law of thermodynamics in a somewhat disguised form; as the statement that heat can only flow (unaided) from a hot body to a cold one. All this gave rise to the field of Classical Thermodynamics, in which the exchange of energy of a macroscopic system in contact with a heat bath was studied. Great thinkers like Clausius, Thomson and Kelvin helped build an extremely fruitful theory in which energy was exchanged both with external forces (work) and with the environment (heat).

In the intervening two centuries, in spite of great strides, many conundrums lingered, largely due to the inability of Classical Thermodynamics to study systems far from equilibrium. Moreover, the requirement of transitioning between states while maintaining the system at near-equilibrium required an infinite time horizon. Thus, it not only failed to describe systems far from equilibrium, but it also erroneously quantified an important characteristic of heat engines: power. On top of that, only a macroscopic description of the thermodynamic system was available.

This would soon change with the advent of Statistical Mechanics, which was able to explain the behavior of equilibrium macroscopic systems in terms of the microscopic motions of their constituent particles. This viewpoint emerged with the development of atomic and molecular theories and represented an extremely satisfying explanation of classical thermodynamics. Nevertheless, it was a theory that concerned both infinite times and big numbers of particles (ensembles), unable to capture the fluctuations that arise from low numbers.

These shortcomings were not completely resolved until the closing of the twentieth century, with the development of Stochastic Thermodynamics [3–6]. Stochastic Thermodynamics studies non-equilibrium systems and their finite time transitions. It provides a framework to quantify notions like heat, work, or entropy

O. Movilla Miangolarra, *Stochastic Thermodynamic Treatment of Thermal Anisotropy*, Springer Theses, https://doi.org/10.1007/978-3-031-68066-3_2

at the level of single particle trajectories. This allows us to study the motion of a single mesoscopic[1] particle in a heat bath, enabling us to quantify the likelihood of large fluctuations, while still allowing us to derive the macroscopic statistics of a system of many such particles. In this way, we are able to study biological systems as well as other non-equilibrium systems and quantify both the passing of time and the power of heat engines.

2.1 Underdamped and Overdamped Systems

Stochastic Thermodynamics gets its name from the fact that the thermodynamic system is modeled by Stochastic Differential Equations (SDEs). Specifically, the system is composed by an ensemble of big particles (mesoscopic Brownian particles) which are immersed in a heat bath of tiny (microscopic) particles (see Fig. 2.1). These small particles are at some temperature T, which is taken to be a scalar throughout this introductory chapter but will be crucially taken as a matrix for the rest of this manuscript. This temperature makes the small particles jitter, thus eventually colliding with the Brownian particle and changing its momentum. This interaction is modeled through "white noise" or, to be more precise, Brownian motion.

The Brownian particle's dynamics are hence dictated by what we call the Langevin equation:

$$dX_t = v_t dt, \tag{2.1a}$$

$$m dv_t = -\nabla U(t, X_t)dt - \gamma v_t dt + \sqrt{2\gamma k_B T} dB_t, \tag{2.1b}$$

which is nothing but Newton's $F = ma$ equation, but for the fact that there is an additional stochastic term accounting for the collisions of these tiny particles with our Brownian particle. Here, $X_t \in \mathbb{R}^n$ denotes the location of a particle,

Fig. 2.1 A Brownian particle (blue), immersed in a heat bath formed by smaller particles at temperature T (red)

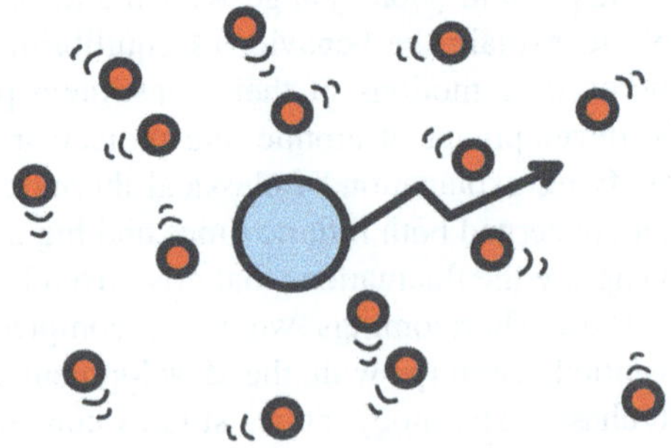

[1] On a scale between microscopic (which is the scale of the heat bath) and macroscopic.

$v_t \in \mathbb{R}^n$ represents its velocity at time t, m its mass, γ the viscosity coefficient of the ambient medium, k_B is the Boltzmann constant and $\{B_t\}$ represents an n-dimensional standard Brownian motion. Furthermore, $U(t, x)$ represents a potential energy that exerts a force $-\nabla U(t, x)$ on a particle at location $x \in \mathbb{R}^n$. This potential constitutes a control action that models the interaction of our ensemble of particles with the external world, where the time t is also thought of as the control parameter.

Equivalently, the dynamics of the Brownian particle (2.1) can be described in terms of their statistics, which are captured in the joint probability density of X_t and v_t, $\rho(t, x, v)$. This means that $\rho(t, x, v)dxdv$ is the probability of finding a particle with a velocity $v_t \in [v, v + dv]$ at position $X_t \in [x, x + dx]$ at time t. This probability density evolves according to the Fokker-Planck equation [7, 8], which for the system in (2.1), takes the form

$$\partial_t \rho = -\nabla_x \cdot (\rho v_t) + \frac{1}{m}[\gamma\rho + (\gamma v_t + \nabla U)\nabla_v \rho] + \frac{\gamma k_B T}{m^2}\Delta_v \rho,$$

dictating the fate of the ensemble of Brownian particles.[2] Unfortunately, this Partial Differential Equation (PDE) is rarely tractable, and therefore it is common to introduce simplifying assumptions.

Specifically, when the inertial effects in the Langevin equation (2.1b) are negligible, i.e. when the time-scales we are interested in are much larger than the time-scale dictating the decay of v_t, $\frac{m}{\gamma}$, we can average out the fast variable v_t to obtain the *overdamped* Langevin equation

$$dX_t = -\frac{1}{\gamma}\nabla U(t, X_t)dt + \sqrt{\frac{2k_B T}{\gamma}}dB_t. \tag{2.2}$$

Formally, the overdamped Langevin equation is obtained from (2.1b) by setting $m = 0$ and replacing $v_t dt = dX_t$. As it is seen in (2.2), the forces acting on the Brownian particle directly affect its position, since in the overdamped setting the velocity degree of freedom does not have any memory; it does not hold any kinetic energy. For a more detailed explanation see [4, page 20]. We will refer to the system in (2.2) as overdamped, and to the one in (2.1), as underdamped.

Similarly, the probability density of X_t, $\rho(t, x)$, evolves according to the overdamped Fokker-Planck equation

$$\partial_t \rho = \frac{1}{\gamma}\nabla \cdot (\rho[\nabla U + k_B T \nabla \log(\rho)]) = -\nabla \cdot (J) = -\nabla \cdot (\rho v), \tag{2.3}$$

[2] Here, ∇_x denotes the gradient with respect to the x variable, i.e. $(\partial_{x_1}, \ldots, \partial_{x_n})'$, and similarly for v, for which we have also introduced the Laplacian $\Delta_v := \nabla_v \cdot (\nabla_v(\cdot))$.

where we have introduced the effective velocity field

$$v(t,x) = -\frac{1}{\gamma}\big(\nabla U(t,x) + k_B T \nabla \log(\rho(t,x))\big), \tag{2.4}$$

as well as the probability current

$$J(t,x) = \rho(t,x)v(t,x),$$

to write the Fokker-Planck equation as a continuity equation.

Overdamped models (2.2) are typical when considering colloidal particle systems and models of biological processes [5, 9, 10]. Given their comparative simplicity and potential for applications, we will restrict ourselves to the study of overdamped systems, to which we will devote the vast majority of this manuscript (the only exception being Appendix A).

2.2 Energetics

Up to now we have successfully defined how our thermodynamic system evolves in time, both at the level of single particle trajectories (2.2), and at the ensemble level (2.3). In order to actually do some thermodynamics, we need to define how the thermodynamic system exchanges energy with both the environment and the external forcing. To this end, let us define the total energy of the overdamped Brownian particle by

$$E = U(t, X_t),$$

where its kinetic energy is neglected since it decays faster than the time scales we are interested in (overdamped assumption). Its incremental change is given by

$$dE = \partial_t U(t, X_t)dt + \nabla U(t, X_t)' \circ dX_t, \tag{2.5}$$

where $'$ denotes transpose and $\circ$ denotes Stratonovich integration (see [7] for a lucid introduction to the subject of stochastic control and stochastic calculus).

Conservation of energy (a.k.a the first law of thermodynamics) implies that this change in energy can only be due to (1) energy exchanged between the system and the environment, or (2) energy exchanged between the system and the external forcing. The former is referred to as heat (denoted by Q), while the latter constitutes work (W). Hence, the first law of thermodynamics at the level of a single particle can be written as

$$dE = đQ + đW, \tag{2.6}$$

where $đ$ denotes that the differential is not exact in that $\int đQ$ and $\int đW$, along a curve, depend on the choice of curve and not only the endpoints.

The energy exchanged between the heat bath and the particle is due to forces[3]

$$-\gamma\frac{dX_t}{dt}+\sqrt{2k_B\gamma T}\frac{dB_t}{dt}$$

applied by the heat bath (see [11, page 19], [12, page 63]). The term $-\gamma\frac{dX_t}{dt}$ is due to dissipation, while $\sqrt{2k_B\gamma T}\frac{dB_t}{dt}$ is due to fluctuations. Thus, the heat at the scale of a single particle is formally expressed as forces times displacement,

$$đQ=\Big(-\gamma\frac{dX_t}{dt}+\sqrt{2k_B\gamma T}\frac{dB_t}{dt}\Big)'\circ dX_t, \tag{2.7}$$

which, in combination with (2.2), takes the precise mathematical form

$$đQ=\nabla U(t,X_t)'\circ dX_t.$$

The incremental work effected through interaction of the particle with the potential U can be written as

$$đW=\partial_t U(t,X_t)dt$$

where we have used the first law of thermodynamics (2.6) and the expression for dE in (2.5). Hence, work relates to how the energy U of our particle changes as a function of our control parameter t. Thus, it captures the energy that flows into the system in a controlled/useful way, while heat accounts for the energy that goes into the system through the rest of the degrees of freedom, which are not under our direct control. It is important to note that the sign convention is chosen such that both work and heat are positive when supplied to the particle.

We can also define these notions at the level of the ensemble by taking the expectation value, denoted by $\mathbb{E}[\cdot]$. In particular, the total energy of the system is given by

$$\mathcal{E}=\mathbb{E}[E]=\int U(t,x)\rho(t,x)dx,$$

where we have used calligraphic letters to denote quantities at the ensemble level. In order to take the expected value of $đQ=\nabla U(t,X_t)'\circ dX_t$, it is convenient to write the Stratonovich differential in Itô form as

[3] Here $\frac{dX_t}{dt}$ is formal (in the sense that it is not meaningful mathematically speaking) and similarly for $\frac{dB_t}{dt}$ which represents white noise.

$$
\begin{aligned}
đQ &= \nabla U(t, X_t)' dX_t + \frac{1}{2} \mathrm{Tr}[\mathrm{Jac}\big(\nabla U(t, X_t)\big) dX_t (dX_t)'] \\
&= \nabla U(t, X_t)' dX_t + \frac{k_B T}{\gamma} \nabla \cdot \big(\nabla U(t, X_t)\big) dt,
\end{aligned}
$$

where Jac(·) denotes the Jacobian, we have used the Itô rule (see [7, Theorem 4.4.2]), and the Langevin equation (2.2) along with the formal Itô identities $dB_t^2 = dt, dB_t dt = 0$, and $dt^2 = 0$. Upon using (2.2) and taking expectation (while keeping in mind that the expected value of $g(t, X_t) dB_t$ is zero for any physical square integrable function[4] g), we obtain the following expression for $đ\mathcal{Q} = \mathbb{E}[đQ]$:

$$
\begin{aligned}
đ\mathcal{Q} &= -\frac{1}{\gamma} \int \|\nabla U(t, x)\|^2 \rho(t, x) dx dt + \frac{k_B T}{\gamma} \int \nabla \cdot \big(\nabla U(t, x)\big) \rho(t, x) dx dt \\
&= -\frac{1}{\gamma} \int \|\nabla U(t, x)\|^2 \rho(t, x) dx dt - \frac{k_B T}{\gamma} \int \nabla U(t, x)' \nabla \rho(t, x) dx dt \\
&= \int \nabla U(t, x)' v(t, x) \rho(t, x) dx dt, \qquad (2.8)
\end{aligned}
$$

where we have used integration by parts (under standard assumptions on the decay rate of ρ at infinity) to obtain the second equality, and the definition of v (2.4) to obtain the third equality. On the other hand, the expected value of the differential of work can be readily obtained to be

$$
đ\mathcal{W} = \mathbb{E}[đW] = \int \partial_t U(t, x) \rho(t, x) dx dt.
$$

Finally, integrating over time we obtain the work and heat transferred to the system during the interval $[0, t_f]$, which can be expressed as

$$
\mathcal{Q} = \int_0^{t_f} \int \langle \nabla U, J \rangle dx dt, \qquad (2.9)
$$

$$
\mathcal{W} = \int_0^{t_f} \int \partial_t U \rho dx dt. \qquad (2.10)
$$

Doing similarly with the energy of the system we get to the first law at the level of an ensemble over a finite-time window:

$$
\Delta \mathcal{E} = \mathcal{W} + \mathcal{Q},
$$

[4] To be precise, $\mathbb{E}[g(t, X_t) dB_t] = 0$ for any adapted function $g(t, X_t)$ such that $\mathbb{E}[\int_0^t \|g(s, X_s)\|^2 ds] < \infty$.

where $\Delta\mathcal{E} = \mathcal{E}|_{t=t_f} - \mathcal{E}|_{t=0} = \int U(t_f, x)\rho(t_f, x)dx - \int U(0, x)\rho(0, x)dx$. We would like to remark that if our system goes over a cycle, we have that $\Delta\mathcal{E} = 0$, and thus,

$$\mathcal{W}_{\text{out}} := -\mathcal{W} = \mathcal{Q},$$

where we have defined the work output $\mathcal{W}_{\text{out}}$.

2.3 Second Law

Having studied the first law of thermodynamics, it is time to derive the second law. The second law of thermodynamics states that the entropy rate of an isolated system is always positive (or more precisely, non-negative). It is a statistical law, in that it applies to ensembles consisting of very large numbers of particles.

In the setting of Stochastic Thermodynamics, the isolated system is composed of both the heat bath and the system itself (consisting of overdamped particles). The entropy rate of the isolated system, to which we refer as total entropy rate, is thus given by the sum of the environment's and system's entropy rates:

$$\dot{\mathcal{S}}_{\text{tot}} = \dot{\mathcal{S}}_{\text{env}} + \dot{\mathcal{S}}_{\text{sys}}.$$

Since the environment is comparatively much larger than the system of particles, it is assumed to be at equilibrium at all times; its thermalization time constant is assumed much faster than other processes taking place (e.g., control actuation). Hence, the entropy rate of the environment can be defined in the classical sense through Clausius equality,

$$\dot{\mathcal{S}}_{\text{env}} = -\frac{\dot{\mathcal{Q}}}{T}, \tag{2.11}$$

where $\dot{\mathcal{Q}} = \frac{d\mathcal{Q}}{dt}$ and the negative sign is due to the heat being positive when flowing from the heat bath to the system.[5] Here, we have crucially used the fact that our system is in contact with a unique heat bath at a unique temperature. As we will see in the following chapters, the need to modify this expression when the system is in simultaneous contact with more than one heat bath will be key in the description of systems under anisotropic fluctuations.

On the other hand, the entropy of the system is given by

$$\mathcal{S}_{\text{sys}} = -k_B \int_{\mathbb{R}^n} \rho \log(\rho) dx,$$

[5] Intuitively, the more heat is transferred to the system of particles, the less energy remains in the environment, and the less "disordered" the environment becomes.

which can be thought of as quantifying both the disorder in our system, as well as the information we have about its current state [13]. The units of entropy are those of the Boltzmann constant (units of energy/temperature), following the standard convention in the physics literature. Taking its time derivative we obtain

$$\dot{\mathcal{S}}_{\text{sys}} = -k_B \int \partial_t \rho \log \rho dx - k_B \cancelto{0}{\int \partial_t \rho dx} \tag{2.12}$$

$$= k_B \int \nabla \cdot (J) \log \rho dx \tag{2.13}$$

$$= -k_B \int \langle \nabla \log \rho, J \rangle dx, \tag{2.14}$$

where the second term in the first equality is zero because the mass of ρ is conserved (equal to one). For the second equality we have used the Fokker-Planck equation (2.3), while for the third equality we have integrated by parts.

Putting (2.11) and (2.12) together, we obtain an expression for the total entropy rate

$$\dot{\mathcal{S}}_{\text{tot}} = \dot{\mathcal{S}}_{\text{env}} + \dot{\mathcal{S}}_{\text{sys}} = -\frac{1}{T} \int \langle \nabla U + k_B T \nabla \log \rho, J \rangle dx = \frac{\gamma}{T} \int \frac{1}{\rho} \|J\|^2 dx,$$

where we have used (2.9) for the first equality, and the definition of the probability current J for the second. This expression is always positive, which constitutes the standard second law. Integrating over the time interval $[0, t_f]$, we obtain

$$\mathcal{S}_{\text{tot}} = \frac{\gamma}{T} \int_0^{t_f} \int \frac{1}{\rho} \|J\|^2 dx dt = \frac{\gamma}{T} \int_0^{t_f} \int \|v\|^2 \rho dx dt. \tag{2.15}$$

We will refer to this expression as entropy production.

As you might have guessed, entropy production is nothing but the cost that we were trying to minimize in the Benamou-Brenier formulation of the optimal mass transport problem, and thereby, it can be bounded from below by $\frac{\gamma}{t_f T} \mathrm{W}_2(\rho(0, \cdot), \rho(t_f, \cdot))^2$. In this way one can obtain the second law that now has a finite time correction:

Second Law

Consider the overdamped Langevin dynamics. Let $\rho_0 = \rho(0, \cdot)$ and $\rho_{t_f} = \rho(t_f, \cdot)$ be the initial and final distributions. Then we have

$$\mathcal{S}_{\text{tot}} \geq \frac{\gamma}{t_f T} \mathrm{W}_2(\rho_0, \rho_{t_f})^2, \tag{2.16}$$

(continued)

where equality is achieved by geodesics in the Wasserstein space, this is, trajectories and velocities that minimize the entropy production.

It is important to remark that these optimal trajectories (geodesics) can be obtained by controlling the velocity v only through the potential forcing ∇U. To see this, note that a selection of ∇U can specify the gradient part of $v = -\frac{1}{\gamma}(\nabla U + k_B T \nabla \log \rho)$, and therefore, any value for $\partial_t \rho$ in (2.3). Specifically, for any $\partial_t \rho = \dot{\rho}$ with $\|\frac{\dot{\rho}}{\rho}\|_\rho^2 < \infty$, the equation

$$\nabla \cdot (\rho \frac{1}{\gamma}(\nabla U + k_B T \nabla \log \rho)) = \dot{\rho}$$

is a Poisson equation and has a (unique) solution for U under assumption A1 on the state ρ (see Sect. 1.4). On the other hand, velocities that minimize entropy production can be attained by controlling the potential forcing ∇U, since optimal velocities are exactly of gradient form as explained in Sect. 1.4. However, as we will see in Chap. 5, this will no longer be the case in the anisotropic setting.

This connection between OMT and Stochastic Thermodynamics [14] has recently attracted a considerable amount of interest and has proven itself extremely fruitful, not only in the design of optimal control protocols both in the continuous (as in here) [15–19] and discrete settings [20, 21], but also in the development of geometric approaches [22–25] and derivation of speed limits and uncertainty relations [25–27]. Part of the goal of this manuscript is to explore its implications in the presence of thermal anisotropy in the environment.

Hence, the second law with finite time corrections implies that, if we drive our system in finite time, positive total entropy production is inevitable. A different, yet equivalent,[6] interpretation of the second law can be found through the notion of free energy. The classical free energy of a state is defined by

$$\mathcal{F} = \mathcal{E} - T\mathcal{S}_{\text{sys}}. \tag{2.17}$$

The second law of thermodynamics provides a bound on the average work required to change the state of a thermodynamic system in terms of the difference between the end-point free energies. Specifically, taking the time derivative of the free energy we obtain

$$\dot{\mathcal{F}} = \dot{\mathcal{E}} - T\dot{\mathcal{S}}_{\text{sys}} = \dot{\mathcal{W}} + \dot{\mathcal{Q}} - T\dot{\mathcal{S}}_{\text{sys}} = \dot{\mathcal{W}} - T\dot{\mathcal{S}}_{\text{tot}}.$$

Integrating over time and using both (2.15) and (2.16),

[6] Equivalent if the process is isothermal and isotropic, as considered throughout this chapter.

$$\mathcal{W} = \Delta\mathcal{F} + \underbrace{\gamma \int_0^{t_f} \int_{\mathbb{R}^n} \|v\|^2 \rho dx dt}_{\mathcal{W}_{\text{diss}}} \geq \Delta\mathcal{F} + \frac{\gamma}{t_f} \mathrm{W}_2(\rho_0, \rho_{t_f})^2, \tag{2.18}$$

where we have defined $\Delta\mathcal{F} = \mathcal{F}|_{t=t_f} - \mathcal{F}|_{t=0}$ and the dissipated work $\mathcal{W}_{\text{diss}}$. Thus, one can think of free energy as the energy available for work extraction in a given state, while dissipation represents the gap between the work that we put into the system and the free energy difference that we actually generate.

2.4 Stochastic Heat Engines

Since Carnot's introduction of ideal heat engines [1], the field of thermodynamics has focused on the conception, modeling and optimization of heat engines. The Carnot and Stirling engines, among others, have been central to the development of the field. Stochastic Thermodynamics has been extensively used to study such heat engines at the microscopic scale, which have been also experimentally realized [9, 28, 29]. Whereas classical thermodynamics works focused on quantifying efficiency, in stochastic thermodynamics finite time allows the study of power. Of course, power comes at the expense of efficiency; intuitively, the faster we go, the more power, while efficiency is typically compromised. Hence, a delicate trade-off between power and efficiency arises.

Let us define power over a cycle as

$$\mathcal{P} = \frac{\mathcal{W}_{\text{out}}}{t_f},$$

while efficiency over a cycle will be defined by [30]

$$\eta = \frac{\mathcal{W}_{\text{out}}}{\mathcal{Q}_{\text{qs}}},$$

where $\mathcal{Q}_{\text{qs}} = \mathcal{W}_{\text{out}} + \mathcal{W}_{\text{diss}}$ is the quasi-static heat (or negative of the quasi-static work over a cycle). Hence, efficiency is defined as the ratio between the work that we get out of a trajectory and the maximum work that we would be able to extract from the same trajectory if we were able to parametrize it in infinite time. Given that $\mathcal{W}_{\text{diss}} \geq 0$, efficiency is bounded by one, while it can turn negative if the work output is negative (if we put work into the system instead of extracting it).

Let us review here the foundational work by Schmiedl and Seifert [31] in which they explore optimal protocols for maximizing work output (and thus in some sense power) of a microscopic Carnot-like engine, this is, an engine that runs in between two alternating temperatures. We will do so from an optimal transport perspective, a similar approach can be found in [15].

Specifically, we consider a one-dimensional overdamped system subject to a quadratic potential:

$$U(t,x) = \frac{1}{2}k(t)x^2.$$

In order to extract work from two heat baths at different temperatures, the overdamped system is subjected to a Carnot cycle that consists of two isothermal steps (constant temperature) and two adiabatic ones (no heat transfer) as follows:

1. Isothermal: The system at a cold temperature (T_c) is compressed by narrowing the quadratic potential,
2. Adiabatic: The system is instantaneously put in contact with a hot temperature (T_h) heat bath while the state of the system is kept constant,
3. Isothermal: The system is expanded by widening a quadratic potential,
4. Adiabatic: The system is instantaneously put in contact with the cold temperature T_c.

This process is depicted in Fig. 2.2.

The work done on the system in each of these steps can be computed, using (2.18) for isothermal steps (with final time $t_f/2$) and $\mathcal{Q} = 0$ in conjunction with the first law for the adiabatic ones, as follows:

1. Isothermal: $\mathcal{W}_1 \geq \Delta U_1 - k_B T_c \Delta \mathcal{S}_{1,\mathrm{sys}} + \frac{2\gamma}{t_f}\mathrm{W}_2(\rho_a, \rho_b)^2 =: \mathcal{W}_1^*$,
2. Adiabatic: $\mathcal{W}_2 = \Delta U_2$,
3. Isothermal: $\mathcal{W}_3 \geq \Delta U_3 - k_B T_h \Delta \mathcal{S}_{3,\mathrm{sys}} + \frac{2\gamma}{t_f}\mathrm{W}_2(\rho_a, \rho_b)^2 =: \mathcal{W}_3^*$,
4. Adiabatic: $\mathcal{W}_4 = \Delta U_4$.

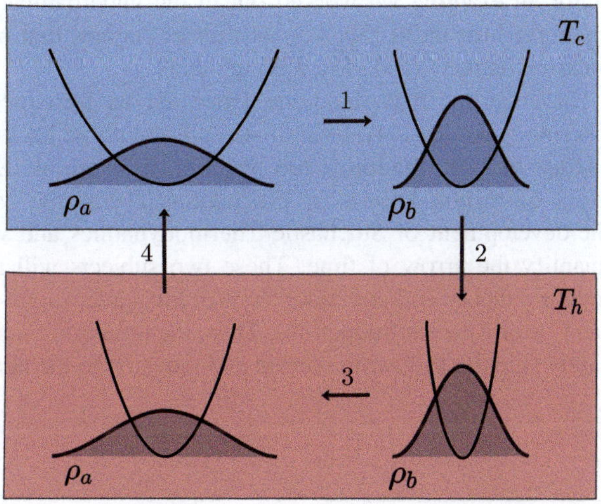

Fig. 2.2 Carnot cycle for stochastic heat engines. Here, the quadratic functions represent the potential, while the filled curves represent the distribution of the particles

Putting all these ingredients together and noting that over a cycle $\Delta\mathcal{E} = \sum_i \Delta U_i = 0$ we obtain that the maximum work output is given by[7]

$$\begin{aligned}\max\{\mathcal{W}_{\text{out}} \mid \rho_a \text{ and } \rho_b \text{ are fixed}\} &= -\mathcal{W}_1^* - \mathcal{W}_2 - \mathcal{W}_3^* - \mathcal{W}_4 \\ &= k_B(T_h - T_c)\Delta\mathcal{S}_{\text{sys}} - \frac{4\gamma}{t_f}\mathrm{W}_2(\rho_a, \rho_b)^2 \\ &=: \mathcal{W}_{\text{out}}^*,\end{aligned}$$

where $\Delta\mathcal{S}_{\text{sys}} = \Delta\mathcal{S}_{3,\text{sys}} = -\Delta\mathcal{S}_{1,\text{sys}} = \mathcal{S}_{\text{sys}}(\rho_a) - \mathcal{S}_{\text{sys}}(\rho_b) > 0$. Therefore, trajectories that maximize work output are Wasserstein-2 geodesics between ρ_a and ρ_b. If ρ_a and ρ_b are Gaussian distributions with zero mean and variance Σ_a and Σ_b, respectively, one can write this expression explicitly in terms of Σ_a and Σ_b using Eq. (1.7), as well as obtain explicit expressions for the optimal trajectories through Eq. (1.6).

Consequently, for any fixed final time t_f, maximum power is given by

$$\mathcal{P}^* = \frac{\mathcal{W}_{\text{out}}^*}{t_f},$$

and the efficiency at maximum power can be written as

$$\eta(\mathcal{P}^*) = 1 - \frac{4\gamma}{t_f}\frac{\mathrm{W}_2(\rho_a, \rho_b)^2}{k_B(T_h - T_c)\Delta\mathcal{S}_{\text{sys}}}.$$

Since the work by Schmiedl and Seifert [31], other engines have been studied in the setting of Stochastic Thermodynamics, such as underdamped engines [32, 33], engines in contact with a unique time-varying heat bath [30, 34–38] (see Appendix A for an example we worked out in the underdamped setting), engines powered by active heat baths [39, 40], as well as engines that optimally extract work from measurements [17, 41–43], among others.

In order to understand the latter, let us introduce the field of Information Thermodynamics [44]. Even if we will not come back to it, the interplay between energy and information is too beautiful to overlook. Similarly, we will introduce a set of results known as Fluctuation Theorems, which have been central in the development of Stochastic Thermodynamics and will help us understand and quantify the arrow of time. These two subjects will rely on certain concepts of measure theory and stochastic control the reader might not be familiar with (again, see [7] for a nice introduction). Thus, the unfamiliar and/or uninterested reader can safely skip the following section and move on to the next chapter.

[7] Throughout, we superscribe $*$ to denote an optimal solution.

2.5 Further Results: Information Thermodynamics and Fluctuation Theorems

2.5.1 *Information Thermodynamics*

In 1871 James Clerk Maxwell devised a demonic creature capable of making the very foundations of Statistical Mechanics—a field himself helped construct—tremble. He pictured a demon in control of a door separating two compartments with gas particles [45]. The demon was capable of measuring the velocities and positions of the particles. She controlled the door such that if a particle with high velocity came from left to right, or a particle with low velocity came from right to left, the door was opened. Otherwise, the door remained closed. This behavior is illustrated in Fig. 2.3. Since, at equilibrium, the equipartition theorem[8] tells us that kinetic energy can be equated with thermal energy, after some time, the demon was able to create a temperature gradient by separating particles with low kinetic energy (left compartment), from those with high kinetic energy (right compartment).

This temperature gradient resulted in an increase of free energy, without paying any work for it, hence contradicting the second law! The apparent paradox was not resolved until, a century later, Rolf Landauer affirmed that information is physical [46]. Starting from the basic assumption that information must be stored somewhere (in this case, information about the position and velocities of the particles), he was able to link the gain of information with the work performed. The relation

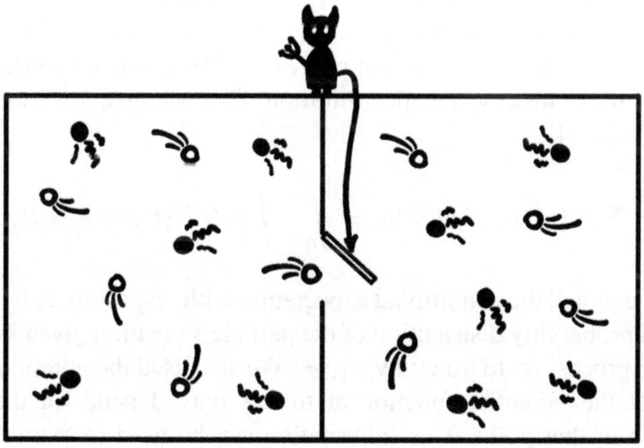

Fig. 2.3 Maxwell's demon. Particles with high velocity (hollow) are allowed to cross from left to right, while particles with low velocity (solid) can go from right to left. This creates a temperature gradient from which work can be extracted

[8] Precisely, the equipartition theorem affirms that, at equilibrium, any degree of freedom which appears only quadratically in the energy has an average energy of $\frac{1}{2}k_B T$.

between information and work gradually became a central theme of Stochastic Thermodynamics [44]. Ideas from stochastic control were naturally brought in and the second law was extended to include discrete time measurements, as well as continuous ones, both for quantum systems [47, 48], and classical systems [49, 50]. In these studies, a generalized version of the second law was introduced:

$$\mathcal{W} \geq \Delta\mathcal{F} - k_B T \mathcal{I}$$

where $\mathcal{I}$ represents the information utilized in effecting a thermodynamic transition. Information engines that work without a temperature gradient and only fueled by information soon followed [41–43].

Herein, we derive a version of this information second law that accounts for finite time corrections. Specifically, we derive tighter forms of the second law for an overdamped system modeled by the Langevin equation and subject to continuous nonlinear measurements. This is based on our work [17], where the study of single (discrete) measurements can also be found. Moreover, in Appendix B we describe, for the interested reader, how to extract work uniquely from the knowledge of the system, this is, we describe an (optimal) information engine.

As of now, we assume to have access to a continuous stream of noisy measurements (Z_t) of the particle's location (X_t, modeled by the overdamped dynamics (2.2)) given by

$$dZ_t = h(X_t)dt + \sigma_v dV_t, \tag{2.19}$$

where $h(\cdot)$ is the observation function, $\{V_t\}$ is a standard Brownian motion representing the noise in measurements, and σ_v is the strength of noise. We assume that $\{V_t\}$ and $\{B_t\}$ are mutually independent processes. The expected work conditioned on the measurement history, i.e. the filtration $\mathcal{Z}_t$ generated by the observation process $\{Z_s; s \in [0, t]\}$, is

$$\mathcal{W}(\mathcal{Z}_{t_f}) = \int_0^{t_f} \mathbb{E}[\partial_t U^{\mathcal{Z}_t}(t, X_t)|\mathcal{Z}_t]dt = \int_0^{t_f} \int \partial_t U^{\mathcal{Z}_t}(t, x)\rho_{X_t|\mathcal{Z}_t}(t, x|\mathcal{Z}_t)dxdt,$$

where we have denoted the conditional expectation with respect to $\mathcal{Z}_t$ by $\mathbb{E}[\cdot|\mathcal{Z}_t]$ and the conditional probability distribution of the particle's position given knowledge of the observation process up to time t by $\rho_{X_t|\mathcal{Z}_t}$. We also used the notation $U^{\mathcal{Z}_t}(t, X_t)$ to indicate that the potential function at time t may depend on the history of observations up to that point. This information can be used to extract work from the system.

The conditional probability distribution $\rho_{X_t|\mathcal{Z}_t}$ evolves according to the Kushner-Stratonovich equation [51]

$$d\rho_{X_t|\mathcal{Z}_t} = -\nabla \cdot (\rho_{X_t|\mathcal{Z}_t} v^{\mathcal{Z}_t})dt + \frac{1}{\sigma_v^2}\rho_{X_t|\mathcal{Z}_t}(h - \hat{h})d\xi_t. \tag{2.20}$$

where

$$v^{\mathcal{Z}_t} = -\frac{1}{\gamma}[\nabla U^{\mathcal{Z}_t} + k_B T \nabla \log(\rho_{X_t|\mathcal{Z}_t})],$$

and $d\xi_t = dZ_t - \hat{h}_t dt$ is the innovation process, with $\hat{h}_t := \mathbb{E}[h(X_t)|\mathcal{Z}_t] = \int h(x)\rho_{X_t|\mathcal{Z}_t}(t, x|\mathcal{Z}_t)dx$. The innovation process thus accounts for the new information that is not captured in our optimal estimate $\hat{h}_t$.

In a similar way as in Sect. 2.3, we can take the derivative of the free energy to obtain a second law-like expression. Recall that the free energy is given by

$$\mathcal{F}(U(t,\cdot), \rho(t,\cdot)) = \int U(t,x)\rho(t,x)dx + k_B T \int \rho(t,x)\log(\rho(t,x))dx.$$

Its derivative, when conditioned on the measurement history, reads

$$\begin{aligned} d\mathcal{F}(U^{\mathcal{Z}_t}, \rho_{X_t|\mathcal{Z}_t}) &= \int \partial_t U^{\mathcal{Z}_t} \rho_{X_t|\mathcal{Z}_t} dx dt + \int \left(U^{\mathcal{Z}_t} + k_B T \log \rho_{X_t|\mathcal{Z}_t}\right) d\rho_{X_t|\mathcal{Z}_t} dx \\ &\quad + \frac{k_B T}{2} \int \frac{(d\rho_{X_t|\mathcal{Z}_t})^2}{\rho_{X_t|\mathcal{Z}_t}} dx \\ &= \left[\int \partial_t U^{\mathcal{Z}_t} \rho_{X_t|\mathcal{Z}_t} dx - \gamma \int \|v^{\mathcal{Z}_t}\|^2 \rho_{X_t|\mathcal{Z}_t} dx \right. \\ &\quad \left. + \frac{k_B T}{2\sigma_v^2} \int \|h - \hat{h}_t\|^2 \rho_{X_t|\mathcal{Z}_t} dx \right] dt \\ &\quad + \frac{1}{\sigma_v^2} \left[\int (U^{\mathcal{Z}_t} + k_B T \log(\rho_{X_t|\mathcal{Z}_t}))\rho_{X_t|\mathcal{Z}_t}(h - \hat{h}) dx \right] d\xi_t, \end{aligned}$$

where the last term from the first equality is the Itô term accounting for the fact that $\rho_{X_t|\mathcal{Z}_t}$ is now a stochastic process, while for the second equality we have used the Kushner-Stratonovich equation (2.20), integration by parts and the definition of $v^{\mathcal{Z}_t}$. Integrating over the interval, taking the expectation over all possible measurements and solving for $\mathcal{W} = \int_0^{t_f} \int \partial_t U^{\mathcal{Z}_t} \rho_{X_t|\mathcal{Z}_t} \rho_{Z_{0:t}} dx dz dt$ yields

$$\begin{aligned} \mathcal{W} = &\int \mathcal{F}(U(t_f,\cdot), \rho_{X_{t_f}|\mathcal{Z}_{t_f}}(t_f, \cdot|z))\rho_{Z_{0:t_f}}(z)dz - \mathcal{F}|_{t=0} \\ &+ \gamma \int_0^{t_f} \int \|v^{\mathcal{Z}_t}(t,x)\|^2 \rho(t,x) dx dt \\ &- \frac{k_B T}{2\sigma_v^2} \int_0^{t_f} \int \|h(x) - \hat{h}_t\|^2 \rho(t,x) dx dt, \end{aligned} \tag{2.21}$$

where $\mathcal{F}|_{t=0} = \mathcal{F}(U(0,\cdot), \rho(0,\cdot))$, $\rho_{Z_{0:t}}(z)$ denotes the probability distribution of a measurement history $z = Z_{0:t}$, we used the fact that ξ_t behaves as a

Brownian motion under conditional expectation [51, Lemma 5.6] and the identity $\rho(t,x) = \int \rho_{X_t|\mathcal{Z}_t}(t,x|z)\rho_{Z_{0:t}}(z)dz$. Note that the last term is nothing but the mutual information between the random processes $X_{0:t_f}$ and $Z_{0:t_f}$ [52],

$$\mathcal{I}(X_{0:t_f}; Z_{0:t_f}) := \frac{1}{2\sigma_v^2}\int_0^{t_f}\int \|h(x) - \hat{h}_t\|^2 \rho(t,x)dxdt. \tag{2.22}$$

On the other hand, the first term on the right-hand-side, the expected free energy at the final time, can be written as

$$\begin{aligned}
&\int \mathcal{F}(U(t_f,\cdot), \rho_{X_{t_f}|\mathcal{Z}_{t_f}}(t_f,\cdot|z))\rho_{Z_{0:t_f}}(z)dz \\
&= \iint U(t_f,x)\rho_{X_{t_f}|\mathcal{Z}_{t_f}}(t_f,x|z)\rho_{Z_{0:t_f}}(z)dxdz \\
&\quad + k_B T \iint \log(\rho_{X_{t_f}|\mathcal{Z}_{t_f}}(t_f,x|z))\rho_{X_{t_f}|\mathcal{Z}_{t_f}}(t_f,x|z)\rho_{Z_{0:t_f}}(z)dxdz \\
&= \int U(t_f,x)\rho(t_f,x)dx - k_B T \mathcal{S}(X_{t_f}|Z_{0:t_f}),
\end{aligned} \tag{2.23}$$

where we have defined the conditional entropy of X_{t_f} given $Z_{0:t_f}$ by

$$\mathcal{S}(X_{t_f}|Z_{0:t_f}) := -\iint \log(\rho_{X_{t_f}|\mathcal{Z}_{t_f}}(t_f,x|z))\rho_{X_{t_f}|\mathcal{Z}_{t_f}}(t_f,x|z)\rho_{Z_{0:t_f}}(z)dxdz.$$

Noting that the mutual information between the random variable X_{t_f} and the random process $Z_{0:t_f}$ is given by

$$\mathcal{I}(X_{t_f}; Z_{0:t_f}) = \mathcal{S}(\rho(t_f,\cdot)) - \mathcal{S}(X_{t_f}|Z_{0:t_f}),$$

where $\mathcal{S}(\rho) = -\int \rho \log \rho dx$ is the Shannon entropy,[9] we can add and subtract $k_B T \mathcal{S}(\rho(t_f,\cdot))$ to obtain

$$\int \mathcal{F}(U(t_f,\cdot), \rho_{X_{t_f}|\mathcal{Z}_{t_f}}(t_f,\cdot|z))\rho_{Z_{0:t_f}}(z)dz = \mathcal{F}|_{t=t_f} + k_B T \mathcal{I}(X_{t_f}; Z_{0:t_f}).$$

Putting this expression and definition (2.22) together with (2.21), we obtain

$$\begin{aligned}
\mathcal{W} &= \Delta\mathcal{F} - k_B T(\mathcal{I}(X_{0:t_f}; Z_{0:t_f}) - \mathcal{I}(X_{t_f}; Z_{0:t_f})) \\
&\quad + \gamma \int_0^{t_f}\int \|v^{\mathcal{Z}_t}(t,x)\|^2 \rho(t,x)dxdt.
\end{aligned}$$

[9] Note that the Shannon entropy is almost identical to the thermodynamic entropy of the system of particles $\mathcal{S}_{\text{sys}}$, differing only in a factor of k_B.

It remains to obtain a lower-bound on the dissipation term. We would like to do so, as in the standard case, by using the Benamou-Brenier formulation of the optimal mass transport problem. However, $v^{\mathcal{Z}_t}(t,x)$ and $\rho(t,x)$ do not satisfy a continuity equation. In order to overcome this difficulty, let us consider

$$\bar{v}(t,x) := \mathbb{E}[v^{\mathcal{Z}_t}(t,X_t)|X_t = x],$$

where the expected value is taken over all measurement histories. Taking the expectation over all measurement histories of both sides of Eq. (2.20), using the identities $\rho(t,x) = \int \rho_{X_t|\mathcal{Z}_t}(t,x|z)\rho_{Z_{0:t}}(z)dz$ and

$$\bar{v}(t,x)\rho(t,x) = \int v^{\mathcal{Z}_t}(t,x)\rho_{X_t|\mathcal{Z}_t}(t,x|z)\rho_{Z_{0:t}}(z)dz,$$

as well as canceling the mean-zero term multiplied by $d\xi_t$, we obtain that the velocity field $\bar{v}(t,x)$ generates the flow for the marginal distribution $\rho(t,x)$, i.e. that $\partial_t\rho = -\nabla\cdot(\rho\bar{v})$.

Hence, we can use the Benamou-Brenier result to write

$$\int_0^{t_f}\int \|\bar{v}(t,x)\|^2\rho(t,x)dxdt \geq \frac{1}{t_f}\mathrm{W}_2(\rho_0,\rho_{t_f})^2,$$

where $\rho_0 = \rho(0,\cdot)$, $\rho_{t_f} = \rho(t_f,\cdot)$ are the initial and final distributions. By Jensen's inequality

$$\mathbb{E}[|v^{\mathcal{Z}_t}(t,X_t)|^2|X_t] \geq |\mathbb{E}[v^{\mathcal{Z}_t}(t,X_t)|X_t]|^2 = |\bar{v}(t,X_t)|^2.$$

Upon taking expectation over X_t and integrating over the time interval,

$$\int_0^{t_f}\mathbb{E}[|v^{\mathcal{Z}_t}(t,X_t)|^2]dt \geq \int_0^{t_f}\mathbb{E}[|\bar{v}(t,X_t)|^2]dt \geq \frac{1}{t_f}\mathrm{W}_2(\rho_0,\rho_{t_f})^2.$$

Thus, we conclude with the second law that now accounts for information:

Information Second Law
Consider the overdamped Langevin dynamics (2.2) coupled to the observation model (2.19). Let $\rho_0 = \rho(0,\cdot)$ and $\rho_{t_f} = \rho(t_f,\cdot)$ be the initial and final distributions. Then we have

$$\mathcal{W} \geq \Delta\mathcal{F} + \frac{\gamma}{t_f}\mathrm{W}_2(\rho_0,\rho_{t_f})^2 - k_BT(\mathcal{I}(X_{0:t_f};Z_{0:t_f}) - \mathcal{I}(X_{t_f};Z_{0:t_f})). \tag{2.24}$$

This second law accounts not only for the extra work we have to put into our system due to finite time (dissipation), but also for the amount of work that we can extract from the system due to information. This is, even if the free energy is not changed, the second law still allows negative work as long as this is greater than the information used. Specifically, we account for two types of information, the mutual information between the particle location and the measurement $\mathcal{I}(X_{0:t_f}; Z_{0:t_f})$, as well as the remaining information $\mathcal{I}(X_{t_f}; Z_{0:t_f})$ that has not been used.

This result provides the first and tightest analysis for the role of information for feedback systems in finite time under continuous nonlinear observation models. Furthermore, in Appendix B we describe an optimal way in which we can extract this extra available work due to measurements, thus creating an engine that runs uniquely due to access to information about the system, in a way akin to Maxwell's demon.

2.5.2 *Fluctuation Theorems*

Next we switch gears and derive some of the most celebrated results in the area of Stochastic Thermodynamics that go under the name of Fluctuation Theorems. The Fluctuation Theorems are comprised of several results that express universal properties of the underlying distribution and whose importance can hardly be overestimated [5]. They were born for deterministic systems [53, 54], and soon extended to stochastic dynamics [55, 56]. A myriad of different results applying to different systems under different assumptions have been developed since. Here, far from attempting a comprehensive review of these results, we will content ourselves with focusing on two of them: Crook's Detailed Fluctuation Theorem [57] and Jarzynski's Equality [58]. We will do so through a stochastic control lens. For a complete introduction to the field, the reader is referred to [6] for a short review and to [59] for a more comprehensive one.

Consider the overdamped system in (2.2), but driven instead by a force $-\nabla U(\lambda(t), X_t)$ where $\lambda(t)$ is a time-dependent control parameter, i.e.

$$dX_t = -\gamma^{-1}\nabla U(\lambda(t), X_t)dt + \sqrt{\frac{2k_B T}{\gamma}}dB_t, \; X_0 \sim \rho(0, \cdot),$$

where X_t is distributed according to $\rho(t, \cdot)$, with $t \in [0, t_f]$. Denote by P_X the probability measure induced by the process X on the space of continuous paths $x = \{x_t : t \in [0, t_f]\}$. Formally, $P_X(x)$ gives the probability that the process X is equal to x (though this probability is zero, and one should replace x with a suitable measurable subset of trajectories). Rigorously, we can identify the density of P_X with respect to the probability measure induced by the process driven only by Brownian motion, as can be seen in what follows.

Denote by P_Z the stationary Wiener measure induced by $dZ_t = \sqrt{\frac{2k_BT}{\gamma}}dB_t$, with Z_0 distributed according to the Lebesgue measure.[10] The Radon-Nikodym derivative of P_X with respect to P_Z (reference measure) is given by Girsanov's theorem,

$$\frac{dP_X}{dP_Z} = \exp\left[-\sum_{i=1}^{n}\left(\frac{1}{2k_BT}\int_0^{t_f} \partial_i U(\lambda(t), X_t)(dX_t)_i \right.\right.$$
$$\left.\left.+\frac{1}{4\gamma k_BT}\int_0^{t_f} |\partial_i U(\lambda(t), X_t)|^2 dt\right)\right]\rho(0,\cdot), \tag{2.25}$$

see [8, Section 3.5]; it represents a probability density with respect to the reference measure. The formal derivation of this expression is fairly simple: express the density for the time discretization of the process and take the continuous-time limit (see Example 4.5.2 in [7]).

Let us define the time-reversal of trajectories and protocols, $\widetilde{x}_t := x_{t_f-t}$ and $\widetilde{\lambda}(t) := \lambda(t_f - t)$, as well as a suitably time-reversed process driven by the same Brownian motion,

$$d\widetilde{X}_t = -\gamma^{-1}\nabla U(\widetilde{\lambda}(t), \widetilde{X}_t)dt + \sqrt{\frac{2k_BT}{\gamma}}dB_t,\ \widetilde{X}_0 \sim \widetilde{\rho}(0,\cdot),$$

where $\widetilde{\rho}(0,\cdot) = \rho(t_f,\cdot)$. Our goal is to compare the probability that $X = x$ with that of $\widetilde{X} = \widetilde{x}$. This can be viewed as performing a Neyman-Pearson type test to validate a hypothesis on the time-arrow.

Indeed, an expression similar to (2.25) can be written for the "backward" law $P_{\widetilde{X}}$. We now wish to evaluate the ratio

$$\frac{dP_X(x)}{dP_{\widetilde{X}}(\widetilde{x})} = \left(\frac{dP_X}{dP_Z}(x)\right)/\left(\frac{dP_{\widetilde{X}}}{dP_Z}(\widetilde{x})\right), \tag{2.26}$$

this is, the ratio between the probability of the forward path x under the forward law, and the probability of the backward path $\widetilde{x}$ under the "backward" law.

Since $\int_0^{t_f} |\partial_i U(\lambda(t), X_t)|^2 dt$ is an ordinary integral, its value is unaffected by time reversal of the integrand, and therefore, setting $X = x$ and $\widetilde{X} = \widetilde{x}$ gives that

$$\int_0^{t_f} |\partial_i U(\lambda(t), x_t)|^2 dt = \int_0^{t_f} |\partial_i U(\widetilde{\lambda}(t), \widetilde{x}_t)|^2 dt$$

---

[10] P_Z is an unbounded positive measure on paths (hence, not a probability measure) and Z_t is referred to as the reversible Wiener process [60].

for all i. On the other hand, when considering stochastic integrals care must be taken. Here, the Itô integral $\int_0^{t_f} \partial_i U(\lambda(t), X_t)(dX_t)_i$ is approximated in mean-square by sums

$$\sum_j \partial_i U(\lambda(t_j), x_{t_j})(x_{t_{j+1}} - x_{t_j}),$$

and likewise, $\int_0^{t_f} \partial_i U(\widetilde{\lambda}(t), \widetilde{X}_t)(d\widetilde{X}_t)_i$ can be approximated at the time-reversed trajectories by

$$\sum_j \partial_i U(\lambda(t_{j+1}), x_{t_{j+1}})(x_{t_j} - x_{t_{j+1}}).$$

Their difference gives

$$\begin{aligned}
&\int_0^{t_f} \partial_i U(\lambda(t), X_t)(dX_t)_i \mid_x - \int_0^{t_f} \partial_i U(\widetilde{\lambda}(t), \widetilde{X}_t)(d\widetilde{X}_t)_i \mid_{\widetilde{x}} \\
&= 2 \int_0^{t_f} \partial_i U(\lambda(t), X_t) \circ (dX_t)_i \mid_x \\
&= 2Q_i.
\end{aligned}$$

where we have used the definition of the Stratonovich integral as the limit of sums

$$\sum_j \frac{\partial_i U(\lambda(t_{j+1}), x_{t_{j+1}}) + \partial_i U(\lambda(t_j), x_{t_j})}{2}(x_{t_{j+1}} - x_{t_j}),$$

and we have defined $Q_i = \int_0^{t_f} \partial_i U(\lambda(t), X_t) \circ (dX_t)_i$.

Therefore, the ratio in (2.26) becomes

$$\frac{dP_X(x)}{dP_{\widetilde{X}}(\widetilde{x})} = \exp\left(-\frac{Q}{k_B T}\right) \frac{\rho(0, x_0)}{\rho(t_f, x_{t_f})}, \tag{2.27}$$

where we have used the fact that $\sum_i Q_i = Q$ denoting the total heat at the level of a single particle trajectory. Much like internal energy, heat, and work that can be defined at the level of individual particles, the entropy of the system too can be defined for an individual particle at X_t within the ensemble as $-k_B \log \rho(t, X_t)$. Thus, the entropy of the system at the level of the ensemble, $-k_B \int \rho(t, x) \log \rho(t, x) dx$, can be interpreted as the mean entropy of particles. Hence, the total entropy production over a transition at the level of a single trajectory is given by $\Delta S_{\text{tot}} = \Delta S_{\text{sys}} + \Delta S_{\text{env}}$, with $\Delta S_{\text{sys}} = -k_B\big(\log\big(\rho(t_f, x_{t_f})\big) - \log\big(\rho(0, x_0)\big)\big)$ and $\Delta S_{\text{env}} = -\frac{Q}{T}$, which allows us to obtain the following statement:

51. Xiong, J.: An Introduction to Stochastic Filtering Theory, vol. 18. Oxford University Press on Demand, Oxford (2008)
52. Duncan, T.E.: On the calculation of mutual information. SIAM J. Appl. Math. **19**(1), 215–220 (1970)
53. Evans, D.J., Cohen, E.G.D., Morriss, G.P.: Probability of second law violations in shearing steady states. Phys. Rev. Lett. **71**, 2401–2404 (1993)
54. Gallavotti, G., Cohen, E.G.D.: Dynamical ensembles in nonequilibrium statistical mechanics. Phys. Rev. Lett. **74** (1995)
55. Kurchan, J.: Fluctuation theorem for stochastic dynamics. J. Phys. A Math. General **31**(16), 3719 (1998)
56. Lebowitz, J.L., Spohn, H.: A gallavotti–cohen-type symmetry in the large deviation functional for stochastic dynamics. J. Stat. Phys. **95**(1), 333–365 (1999)
57. Crooks, G.E.: Entropy production fluctuation theorem and the nonequilibrium work relation for free energy differences. Phys. Rev. E **60** (1999)
58. Jarzynski, C.: Nonequilibrium equality for free energy differences. Phys. Rev. Lett. **78** (1996)
59. Schuster, H.G.: Nonequilibrium Statistical Physics of Small Systems: Fluctuation Relations and Beyond. John Wiley & Sons, New York (2013)
60. Léonard, C.: Some properties of path measures. Séminaire de Probabilités XLVI, pp. 207–230 (2014)

Chapter 3
Stochastic Thermodynamic Systems Subject to Anisotropic Fluctuations

Life on Earth is possible thanks to the temperature difference between the hot Sun and the cold stary sky. This difference provides "negative entropy" that organisms and complex biochemical processes feed upon [1]. Indeed, the Sun provides photons at around 6000 Kelvin that are absorbed and then re-emitted back to the cosmos at about 300 Kelvin, a 20-fold decrease. The "hard currency" paid along the way is a positive entropy rate for the universe as a whole, while at the same time, the anisotropy in the thermal environment (created by the hot and cold temperature heat baths, the sun and the sky, respectively) powers biological engines that make life possible [2, 3].

Whether anisotropic fluctuations come from temperature gradients, concentration gradients across cell membranes or non-equilibrium heat baths [4], the study of thermodynamic systems under anisotropic fluctuations is indeed of central importance in our venture to understand life. The goal of the present chapter is to introduce these thermodynamic systems and their energetics, paving the way to the design of optimal protocols for maximizing work output or minimizing entropy production, which will be explored in Chaps. 4 and 5, respectively. Specifically, we introduce an overdamped thermodynamic system operating far from equilibrium, in which each degree of freedom is in contact with a different temperature heat bath.

3.1 Overdamped Anisotropic Systems

While conventional overdamped systems, as the ones we have seen up until now, represent a collection of particles in contact with a single heat bath at any time, we consider a more general setting, where the thermodynamic system has n coupled degrees of freedom which are subject to fluctuations of different intensities; that is, the system is in contact with multiple heat baths at the same time. Thus, our basic model is the following Langevin system of stochastic differential equations

O. Movilla Miangolarra, *Stochastic Thermodynamic Treatment of Thermal Anisotropy*, Springer Theses, https://doi.org/10.1007/978-3-031-68066-3_3

$$dX_t = \frac{1}{\gamma}F(t, X_t)dt + \sqrt{\frac{2k_B T}{\gamma}}dB_t, \tag{3.1}$$

where the temperature,

$$T = \text{diag}\,(T_1, T_2, \ldots, T_n),$$

is now a diagonal matrix with the value of temperature along the specified n degrees of freedom as entries. Moreover, $F(t, X_t)$ represents an external force that accounts for both conservative and non-conservative actuation, this is, it is given by

$$F(t, x) = -\nabla U(t, x) + f(t, x),$$

where $-\nabla U$ is a potential force (conservative) as before, while f represents non-conservative forcing, this is, forcing that does not come from the gradient of a potential energy. This generalization of the forces that the system is subjected to will be crucial in our study of anisotropic systems, as it will soon become apparent. The degrees of freedom in (3.1) may represent voltages in an electrical circuit with resistors subject to different Johnson-Nyquist thermal noise [5], or the location of a single particle subject to radiations of different intensities in different directions [6].

Similarly to before, the state of the thermodynamic system is represented by the probability density function $\rho(t, x)$ that satisfies the Fokker-Planck equation

$$\partial_t \rho(t, x) + \nabla \cdot J(t, x) = 0, \tag{3.2}$$

where the probability current J is given by

$$\begin{aligned} J(t, x) &= -\frac{1}{\gamma}\big(\nabla U(t, x) - f(t, x) + k_B T \nabla \log \rho(t, x)\big)\rho(t, x) \\ &= v(t, x)\rho(t, x). \end{aligned} \tag{3.3}$$

3.2 Energetics

During thermodynamic transitions, energy is continuously exchanged between particles, external actuation and the surrounding heat baths through work and heat. Once again, the incremental change in internal energy $E_t = U(t, X_t)$ of a single particle at location X_t can be expressed as

$$dE_t = \partial_t U(t, X_t)dt + \nabla U(t, X_t)' \circ dX_t, \tag{3.4}$$

where the non-conservative force f plays no role.

On the other hand, the incremental heat at the level of a single particle (following the same argument as in (2.7)) now contains an extra term due to f,

$$đQ = (\nabla U(t, X_t) - f(t, X_t))' \circ dX_t, \tag{3.5}$$

and can be written in terms of the contributions from the different heat baths as

$$đQ = \sum_i đQ_i = \sum_i (\partial_i U(t, X_t) - f_i(t, X_t)) \circ (dX_t)_i. \tag{3.6}$$

Upon using the Itô rule to express this as an Itô differential and taking the expectation (in the exact same way as we did in (2.8)), the heat rate that flows into the thermodynamic system from the i-th reservoir at the level of the ensemble becomes

$$đ\mathcal{Q}_i = \int (\partial_i U(t, x) - f_i(t, x)) J_i(t, x) dx dt. \tag{3.7}$$

Here, the influx of heat from the i-th reservoir takes place along the i-th degree of freedom, which is coupled to the rest via the potential U and the force f.

In turn, the incremental work can be written as

$$đW = \partial_t U(t, X_t) dt + f(t, X_t)' \circ dX_t,$$

where we have used the expressions for dE and $đQ$, Eqs. (3.4) and (3.5). Following again the same procedure as in (2.8) (expressing the Stratonovich differential in Itô form and taking expectation), we obtain the work at the level of an ensemble to be

$$đ\mathcal{W} = \int \partial_t U \rho dx dt + \int \langle f, J \rangle dx dt.$$

3.3 Second Law

The change in total entropy includes two terms, entropy change within the system and entropy change in the environment,

$$\dot{\mathcal{S}}_{\text{tot}} = \dot{\mathcal{S}}_{\text{sys}} + \dot{\mathcal{S}}_{\text{env}}.$$

The entropy of the system $\mathcal{S}_{\text{sys}} = -k_B \int \log(\rho) \rho dx$ changes with rate,

$$\dot{\mathcal{S}}_{\text{sys}} = -k_B \int \log(\rho) \partial_t \rho \, dx = -k_B \int \langle \nabla \log \rho, J \rangle dx \tag{3.8}$$

where the second equality follows from $\partial_t \rho = -\nabla \cdot J$ and integration by parts. The entropy of the environment changes due to heat exchange according to

$$\dot{\mathcal{S}}_{\text{env}} = -\sum_i \frac{\dot{Q}_i}{T_i} = -\int \langle \nabla U - f, J \rangle_{T^{-1}} dx, \tag{3.9}$$

where we have used (3.7) and defined the inner product $\langle v_1, v_2 \rangle_M = v_1' M v_2$ for some symmetric positive definite matrix M.

Together (3.8) and (3.9) give the total entropy production rate

$$\dot{\mathcal{S}}_{\text{tot}} = -\int \langle T^{-1} \left(\nabla U - f + k_B T \nabla \log \rho \right), J \rangle dx = \int \frac{1}{\rho} \| J \|^2_{\gamma T^{-1}} dx,$$

where we have used the definition of J in (3.3) and the norm induced by the $\langle \cdot, \cdot \rangle_{\gamma T^{-1}}$ inner-product, $\| \cdot \|_{\gamma T^{-1}}$. Therefore, a non-vanishing probability current J irreversibly increases the total entropy; this constitutes the second law of thermodynamics, i.e. $\mathcal{S}_{\text{tot}} \geq 0$. Alternatively, the entropy production rate expressed in terms of v is

$$\dot{\mathcal{S}}_{\text{tot}} = \gamma \int \rho \| v \|^2_{T^{-1}} dx. \tag{3.10}$$

We are now in position to understand a salient feature of anisotropic systems: the system may be at steady state, this is, $\partial_t \rho = 0 = -\nabla \cdot J$, while J need not be zero. Specifically, $\nabla \cdot J = 0$ implies that the orthogonal part of $J = -\frac{1}{\gamma}(\nabla U - f + k_B T \nabla \log \rho)\rho$ to divergence-free vector fields with respect to ρ (a.k.a. the gradient part) vanishes. However, both the temperature being a matrix and f being non-zero introduce divergence-free terms that allow for $\nabla \cdot J = 0$ while $J \neq 0$. Such states are known as *non-equilibrium steady-states* (NESS for short). They are characterized by the circulation present in the steady-state probability current that mediates heat transfer between thermal baths (see (2.8)), maintaining the system far from equilibrium at the cost of a positive entropy production rate at the environment (and also in (3.10)). The condition $J = 0$, where entropy production vanishes, is referred to as *detailed balance* or *micro-canonical reversibility*, and the state, as an equilibrium state.

3.3.1 Fluctuation Theorems

Consider a single random realization $\{X_t \in \mathbb{R}^n;\ t \in [0, t_f]\}$, interpreted as the trajectory of a particle that is an element of an ensemble distributed according to $\rho(t, x)$. The change in the entropy of the system, at the level of a single trajectory, is

$$\Delta S_{\text{sys}} = -k_B \log \rho(t_f, X_{t_f}) + k_B \log \rho(0, X_0)$$
$$= -\int_0^{t_f} k_B \nabla \log(\rho(t, X_t))' \circ dX_t - \int_0^{t_f} k_B \frac{\partial_t \rho(t, X_t)}{\rho(t, X_t)} dt,$$

where we have used the Itô rule on $d(-k_B \log \rho(t, X_t))$. Similarly, the change in the entropy of the environment is

$$\Delta S_{\text{env}} = -\int_0^{t_f} \sum_{i=1}^{n} \frac{đQ_i}{T_i} = -\int_0^{t_f} (T^{-1}\nabla U(t, X_t) - T^{-1} f(t, X_t))' \circ dX_t,$$

leading to the total entropy production for a single trajectory

$$\Delta S_{\text{tot}} = \Delta S_{\text{env}} + \Delta S_{\text{sys}} = \int_0^{t_f} \gamma T^{-1} v(t, X_t)' \circ dX_t + k_B \int_0^{t_f} \frac{\nabla \cdot J(t, X_t)}{\rho(t, X_t)} dt,$$

where we have used the definition of v (3.3) and the Fokker-Planck equation (3.2).

Following the same proof as in Sect. 2.5.2 while carefully accounting for the different temperatures (change $T \to T_i$ in (2.25)), we can quantify the likelihood of fluctuations of ΔS_{tot} through the following Fluctuation Theorems:

Fluctuation Theorems for Thermally Anisotropic Systems

Let X, x denote a forward process and a particular trajectory, while $\widetilde{X}$, $\widetilde{x}$ denote their time reversed counterparts. Then the ratio of the forward probability of the forward trajectory and the "backward" probability of the backward trajectory is

$$\frac{dP_X(x)}{dP_{\widetilde{X}}(\widetilde{x})} = \exp\left(\frac{\Delta S_{\text{tot}}}{k_B}\right). \tag{3.11}$$

Taking the expected value of the inverse of this expression we arrive at

$$\mathbb{E}\left[\exp\left(-\frac{\Delta S_{\text{tot}}}{k_B}\right)\right] = 1. \tag{3.12}$$

Note that if the end-point distributions are equilibrium distributions, we can write the last identity in terms of equilibrium free energy differences, similarly to (2.30). However, this is no longer the case if the end-point distributions are non-equilibrium steady-states. Indeed, in the presence of anisotropy it is not clear how to meaningfully define a free energy-like quantity for general states, a point that is briefly touched upon in Chap. 4.

3.4 A Canonical Example: The Brownian Gyrator

We now specialize to the case where the system is two-dimensional and the forcing is uniquely due to a quadratic potential, namely, $f = 0$ and $U(t, x) = x'K(t)x/2$ with $x \in \mathbb{R}^2$ and $K(t) = K(t)'$ a symmetric matrix. Starting from an initial zero-mean Gaussian distribution, the thermodynamic state traces a path on the submanifold of Gaussian distributions

$$\rho(t, x) = \frac{1}{2\pi \det(\Sigma(t))^{1/2}} e^{-\frac{1}{2}\|x\|^2_{\Sigma(t)^{-1}}}, \tag{3.13}$$

where $\Sigma(t) = \mathbb{E}[X_t X_t']$ is the covariance matrix.

These zero-mean Gaussian paths can be uniquely determined by $\{\Sigma(t) : t \in [0, t_f]\}$, where $\Sigma(t)$ satisfies the differential Lyapunov equation

$$\gamma \dot{\Sigma}(t) = -K(t)\Sigma(t) - \Sigma(t)K(t) + 2k_B T, \tag{3.14}$$

with $T = \mathrm{diag}(T_1, T_2)$ and $T_1 > T_2$. This equation can be obtained by writing a stochastic differential equation for the tensor product $X_t X_t'$ through the Itô rule, i.e.,

$$\begin{aligned} d(X_t X_t') &= (dX_t)X_t' + X_t(dX_t)' + (dX_t)(dX_t)' \\ &= -\frac{1}{\gamma}K(t)X_t X_t' dt - \frac{1}{\gamma}X_t X_t' K(t)dt + \sqrt{\frac{2k_B T}{\gamma}} dB_t X_t' \\ &\quad + X_t\left(\sqrt{\frac{2k_B T}{\gamma}} dB_t\right)' + \frac{2k_B T}{\gamma}dt, \end{aligned}$$

where we have used the overdamped Langevin equation (3.1) along with the formal Itô identities $dB_t^2 = dt$, $dB_t dt = 0$, and $dt^2 = 0$. Taking the expectation, the stochastic terms disappear and the differential Lyapunov equation (3.14) follows.

For this Gaussian case, the velocity field in (3.3) takes form

$$v(t, x) = -\frac{1}{\gamma}(K(t) - k_B T \Sigma(t)^{-1})x,$$

linear in x, since $\nabla \log \rho(t, x) = -\Sigma(t)^{-1}x$.

3.4.1 Steady-State Analysis

When the potential remains constant, with $K(t) = K_c$ positive definite, the system reaches a steady-state distribution (i.e. $\dot{\Sigma}(t) = 0$), which is Gaussian[1] $\mathcal{N}(0, \Sigma_{ss})$ with the steady-state covariance Σ_{ss} satisfying the algebraic Lyapunov equation

$$K_c \Sigma_{ss} + \Sigma_{ss} K_c = 2k_B T. \tag{3.15}$$

The solution to (3.15) is unique and can be expressed as

$$\Sigma_{ss} = 2k_B \int_0^\infty e^{-\tau K_c} T e^{-\tau K_c} d\tau = L_{K_c}(2k_B T), \tag{3.16}$$

where

$$X \mapsto L_A(X) := \int_0^\infty e^{-\tau A} X e^{-\tau A} d\tau,$$

is a linear operator that depends on a symmetric positive definite matrix A.

Then, the velocity field can be written as

$$v(x) = -\frac{1}{\gamma}(K_c - k_B T \Sigma_{ss}^{-1})x = -\frac{1}{2\gamma}(K_c \Sigma_{ss} - \Sigma_{ss} K_c)\Sigma_{ss}^{-1} x, \tag{3.17}$$

where for the second equality we have used the algebraic Lyapunov equation (3.15). It is seen that in order to ensure detailed balance ($J = 0$), K_c and Σ_{ss} must commute. In view of (3.15), also (3.16), this only happens when T and K_c commute; i.e., this happens when K_c is diagonal, since T is already diagonal. In that case, $\Sigma_{ss} = k_B T K_c^{-1}$ is also diagonal and results in zero probability current according to (3.17). Thus, heat cannot transfer between the degrees of freedom and the system is at equilibrium. On the other hand, when K_c and T do not commute, detailed balance breaks down and a non-vanishing (stationary) probability current materializes leading to a non-equilibrium steady-state with non-vanishing heat transfer between the heat baths.

The system described here is known as the Brownian Gyrator. It has been thoroughly studied, since its conception by Filliger and Reimann in 2007 [7], especially at steady-state. In particular, most works have focused on the curl-carrying probability current that mediates a heat transfer between heat baths and the torque that it generates [5, 7–10]. Other works have studied maximum work extraction through periodic variation of the potential function [11, 12] (which we will discuss in detail in Chap. 4), entropy production [13, 14] (Chap. 5), optimal transitioning between states [15], the role of information flow [16, 17], the effect

[1] $\mathcal{N}(0, \Sigma_{ss})$ denotes a Normal (a.k.a Gaussian) distribution with zero mean and covariance Σ_{ss}.

of external forces [18], strong coupling limits [19], the control relevance of anharmonic potentials [20], gyration for underdamped mesoscopic systems [21, 22], the relevance of non-Markovian noise [23] and the use of active reservoirs [24]. Similar systems had been previously introduced in the context of Feynman Ratchets [25], e.g. [26–29], where the potential forcing is not necessarily quadratic.

Herein, we summarize results regarding the heat transferred at steady state and the consequent torque. Using (3.7) together with (3.17), the heat transferred from the first heat bath to the system (denoted by $\mathcal{Q}_1$) can be written as

$$\begin{aligned} đ\mathcal{Q}_1 &= -\frac{1}{2\gamma}\mathbb{E}\Big[(K_cX_t)_1\Big((K_c\Sigma_{ss} - \Sigma_{ss}K_c)\Sigma_{ss}^{-1}X_t\Big)_1\Big]dt \\ &= -\frac{1}{2\gamma}\Big(K_c(K_c\Sigma_{ss} - \Sigma_{ss}K_c)\Big)_{11}dt, \end{aligned}$$

where $(M)_{ij}$ denotes the i, j-th component of matrix M. An expression for $đ\mathcal{Q}_2$ can be obtained by simply switching $1 \leftrightarrow 2$. The first law also implies that $đ\mathcal{Q}_1 = -đ\mathcal{Q}_2$, since at steady-state there is no work performed or energy change. Thus, there is a continuous heat flow through the system; heat is absorbed from one heat bath and released into the other, with no net heat into the system. Note that both $đ\mathcal{Q}_1$ and $đ\mathcal{Q}_2$ only vanish if K_c, and thus Σ_{ss}, are diagonal. Hence, even if at steady-state, this heat transfer leads to non-vanishing entropy production at the heat baths, namely,

$$\dot{\mathcal{S}}_{\text{tot}} = \dot{\mathcal{S}}_{\text{env}} = -\frac{\dot{\mathcal{Q}}_1}{T_1} - \frac{\dot{\mathcal{Q}}_2}{T_2} = \dot{\mathcal{Q}}_1\left(\frac{1}{T_2} - \frac{1}{T_1}\right).$$

On the other hand, following [7], the average torque exerted on the potential can be defined through

$$M = \mathbb{E}[X_1\partial_2U(t, X_1, X_2) - X_2\partial_1U(t, X_1, X_2)]. \tag{3.18}$$

Using the quadratic expression for the potential U, we obtain

$$\begin{aligned} M &= K_{21}\Sigma_{11} + K_{22}\Sigma_{12} - (K_{11}\Sigma_{12} + K_{12}\Sigma_{22}) \\ &= \text{Tr}\Big[K_c\Sigma_{ss}\hat{\Omega}\Big] \\ &= \frac{1}{2}\text{Tr}\Big[(K_c\Sigma_{ss} - \Sigma_{ss}K_c)\hat{\Omega}\Big], \end{aligned}$$

where we have defined the skew-symmetric matrix

$$\hat{\Omega} = \begin{bmatrix} 0 & 1 \\ -1 & 0 \end{bmatrix}.$$

One can now use the expression for Σ_{ss} in (3.16) to write the torque (and also the heat) as an explicit function of T_1, T_2 and the components of K_c only. Instead of doing so (which is straightforward but lengthy; see (4.11) for a similar computation), we will content ourselves with pointing out that the torque is zero if and only if K_c and Σ_{ss} commute. For (more) explicit expressions of the torque and the heat the reader is referred to [7] or [10].

3.4.2 Experimental Realizations

This two-dimensional steady-state system subject to anisotropic temperatures has been experimentally realized in different physical embodiments [5, 8, 10, 30, 31].

The first experimental realization of the Brownian gyrator was introduced by Ciliberto et al. in 2013 [30]. Therein, the degrees of freedom in (3.1) represent charges at different capacitors. Specifically, they considered the circuit depicted in Fig. 3.1a, which consists of three capacitors and two resistors. The resistors are in contact with heat baths of different temperatures, which generate Johnson-Nyquist fluctuating currents. Thus, the state of this electrical-thermal system comprises of the charges $q_1(t)$ and $q_2(t)$ at capacitances C_1 and C_2. Let

$$q_t = \begin{bmatrix} q_1(t) \\ q_2(t) \end{bmatrix}, \quad R = \begin{bmatrix} R_1 & 0 \\ 0 & R_2 \end{bmatrix} \quad \text{and} \quad C = \begin{bmatrix} C_1 + C_c & -C_c \\ -C_c & C_2 + C_c \end{bmatrix},$$

denote the charge vector, and the resistance and capacitance matrices, respectively. Using standard circuit analysis, we find that the equations governing the dynamics of these charges are given by [5]

$$dq_t = -R^{-1}C^{-1}q_t dt + R^{-1}D dB_t, \tag{3.19}$$

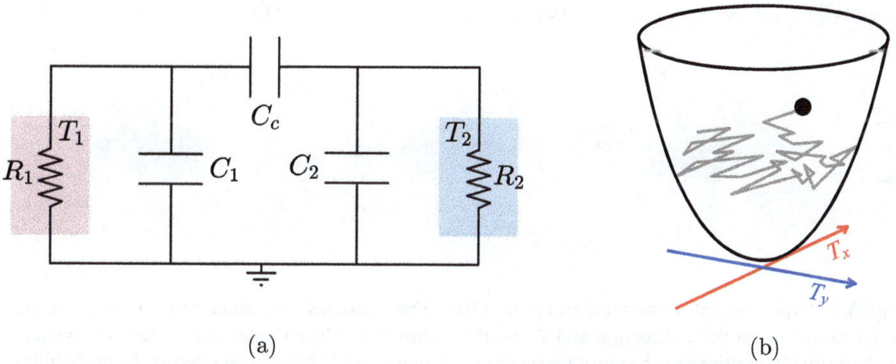

Fig. 3.1 Physical embodiments of the Brownian Gyrator. (**a**) Electrical embodiment of the Brownian Gyrator in which the charges at the capacitors oscillate due to the Johnson-Nyquist noise at the resistors (image from [33]). (**b**) A particle suspended on a quadratic potential is subjected to different fluctuations in different directions

where $\{B_t\}$ models Johnson-Nyquist noise $R^{-1}DdB_t$ at the two resistors [32], with

$$D = \begin{bmatrix} \sqrt{2k_B R_1 T_1} & 0 \\ 0 & \sqrt{2k_B R_2 T_2} \end{bmatrix},$$

with T_1 and T_2 the temperature to which the corresponding resistors are subjected. Let $U(q) = \frac{1}{2}q'C^{-1}q$ be the (potential) energy in the system of capacitances (C_1, C_2, C_c). Then, the state Eq. (3.19) becomes

$$dq_t = -R^{-1}\nabla U(q)dt + R^{-1}DdB_t,$$

which is nothing but a two-dimensional overdamped Langevin equation (3.1) in the absence of non-conservative forces, with R playing the role of the friction coefficient[2] γ.

Perhaps a more intuitive experimental realization of the Brownian Gyrator was introduced later in [10], where a (charged) particle in two dimensions is subjected to a quadratic potential through optical tweezers (see Fig. 3.1b). The anisotropy of the fluctuations is artificially generated by a randomly oscillating electric field in one of the directions. Figure 3.2 displays the experimental results obtained in [10].

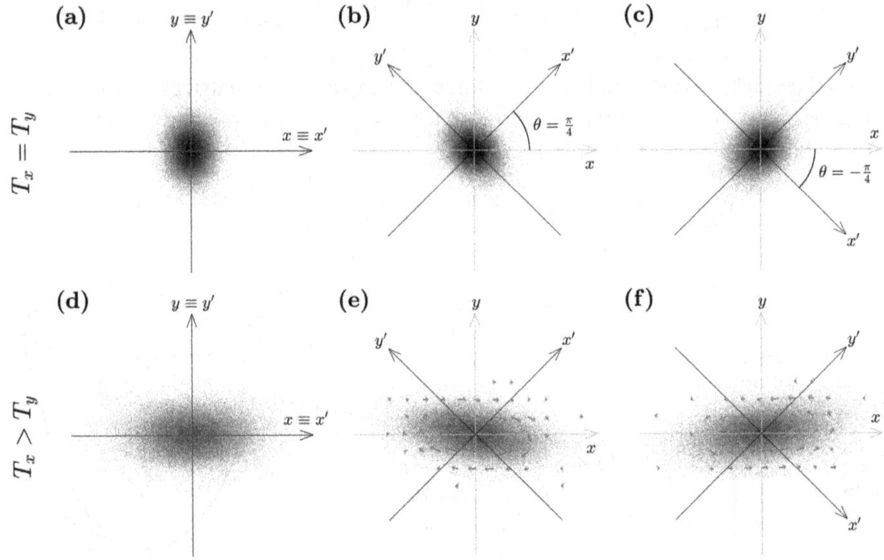

Fig. 3.2 Experimental results obtained in [10]. The particles are subjected to an effective temperature T_x on the x direction and T_y on the y direction. Grey shades denote the steady-state probability distributions of the particles on the $x - y$ plane, while blue arrows depict the probability current. The $x' - y'$ axes correspond to the principal axes of the quadratic potential

[2] Here, R is a diagonal matrix with possibly unequal entries. However, throughout this manuscript we restrict ourselves to the study of systems with scalar friction coefficient γ, and thus for all practical purposes it will suffice to consider $R_1 = R_2$.

It is observed that when the two temperatures are equal (isotropic case: (a), (b), (c)), there is no probability current, regardless of whether the potential principal axes (x' and y') are aligned with the temperature axes (x and y) or not—i.e. regardless of whether K_c is diagonal or not. However, when one temperature is greater than the other ($T_x > T_y$ in (d), (e), (f)) a probability current of swirling particles arises (represented by blue arrows) as long as the principal axes of the potential and those of the temperature are misaligned (i.e. as long as Σ_{ss} and K_c do not commute).

Having introduced anisotropic systems and, in particular, the steady-state circulating characteristics of the Brownian Gyrator, we are now in position to take the next natural step and use the anisotropy of the environment as a means of extracting energy from the system. This will be the content of the following chapter.

References

1. Opatrný, T., Richterek, L., Bakala, P.: Life under a black sun. Am. J. Phys. (2017)
2. Battle, C., Broedersz, C.P., Fakhri, N., Geyer, V.F., Howard, J., Schmidt, C.F., MacKintosh, F.C.: Broken detailed balance at mesoscopic scales in active biological systems. Science (2016)
3. Gnesotto, F.S., Mura, F., Gladrow, J., Broedersz, C.P.: Broken detailed balance and non-equilibrium dynamics in living systems: a review. Rep. Progress Phys. (2018)
4. Puglisi, A., Marconi, U.M.B.: Clausius relation for active particles: what can we learn from fluctuations. Entropy **19**(7), 356 (2017)
5. Chiang, K.-H., Lee, C.-L., Lai, P.-Y., Chen, Y.-F.: Electrical autonomous Brownian gyrator. Phys. Rev. E **96**, 032123 (2017)
6. Bang, J., Pan, R., Hoang, T.M., Ahn, J., Jarzynski, C., Quan, H.T., Li, T.: Experimental realization of feynman's ratchet. New J. Phys. **20**(10), 103032 (2018)
7. Filliger, R., Reimann, P.: Brownian gyrator: a minimal heat engine on the nanoscale. Phys. Rev. Lett. **99**, 230602 (2007)
8. Ciliberto, S., Imparato, A., Naert, A., Tanase, M.: Statistical properties of the energy exchanged between two heat baths coupled by thermal fluctuations. J. Stat. Mech. Theory Exp. **2013**(12), P12014 (2013)
9. Dotsenko, V., Maciołek, A., Vasilyev, O., Oshanin, G.: Two-temperature Langevin dynamics in a parabolic potential. Phys. Rev. E **87**, 062130 (2013)
10. Argun, A., Soni, J., Dabelow, L., Bo, S., Pesce, G., Eichhorn, R., Volpe, G.: Experimental realization of a minimal microscopic heat engine. Phys. Rev. E **96**, 052106 (2017)
11. Movilla Miangolarra, O., Taghvaei, A., Fu, R., Chen, Y., Georgiou, T.T.: Energy harvesting from anisotropic fluctuations. Phys. Rev. E (2021)
12. Movilla Miangolarra, O., Taghvaei, A., Chen, Y., Georgiou, T.T.: Geometry of finite-time thermodynamic cycles with anisotropic thermal fluctuations. IEEE Control Syst. Lett. **6**, 3409–3414 (2022)
13. Movilla Miangolarra, O., Taghvaei, A., Georgiou, T.T.: Minimal entropy production in anisotropic temperature fields. In: 2023 American Control Conference (ACC), pp. 616–621. IEEE (2023)
14. Movilla Miangolarra, O., Taghvaei, A., Georgiou, T.T.: Minimal entropy production in the presence of anisotropic fluctuations. In: IEEE Transactions on Automatic Control (2025)
15. Baldassarri, A., Puglisi, A., Sesta, L.: Engineered swift equilibration of a Brownian gyrator. Phys. Rev. E **102**(3), 030105 (2020)
16. Allahverdyan, A., Janzing, D., Mahler, G.: Thermodynamic efficiency of information and heat flow. J. Stat. Mech. Theory Exp. **2009** (2009)

17. Loos, S.A.M., Klapp, S.H.L.: Irreversibility, heat and information flows induced by non-reciprocal interactions. New J. Phys. **22**(12), 123051 (2020)
18. Cerasoli, S., Dotsenko, V., Oshanin, G., Rondoni, L.: Asymmetry relations and effective temperatures for biased brownian gyrators. Phys. Rev. E **98**, 042149 (2018)
19. Fogedby, H.C., Imparato, A.: A minimal model of an autonomous thermal motor. Europhys. Lett. **119** (2017)
20. Chang, H., Lee, C.-L., Lai, P.-Y., Chen, Y.-F.: Autonomous brownian gyrators: a study on gyrating characteristics. Phys. Rev. E **103**, 022128 (2021)
21. Chun, H.-M., Noh, J.D.: Hidden entropy production by fast variables. Phys. Rev. E **91**(5), 052128 (2015)
22. Bae, Y., Lee, S., Kim, J., Jeong, H.: Inertial effects on the brownian gyrator. Phys. Rev. E **103**, 032148 (2021)
23. Nascimento, E.D.S., Morgado, W.A.M.: Stationary properties of a non-markovian brownian gyrator. J. Stat. Mech. Theory Exp. **2021**(1), 013301 (2021)
24. Lee, J.S., Park, J.-M., Park, H.: Brownian heat engine with active reservoirs. Phys. Rev. E **102**, 032116 (2020)
25. Feynman, R.P., Leighton, R.B., Sands, M.: The Feynman Lectures on Physics, vol. 1. Addison-Wesley Publishing Co., Reading (1963)
26. Magnasco, M.O., Stolovitzky, G.: Feynman's ratchet and pawl. J. Stat. Phys. **93**, 615–632 (1998)
27. Sekimoto, K.: Kinetic characterization of heat bath and the energetics of thermal ratchet models. J. Phys. Soc. Jpn. **66**(5), 1234–1237 (1997)
28. Van den Broeck, C., Kawai, R., Meurs, P.: Microscopic analysis of a thermal brownian motor. Phys. Rev. Lett. **93**(9), 090601 (2004)
29. Gomez-Marin, A., Sancho, J.M.: Heat fluctuations in brownian transducers. Phys. Rev. E **73**(4), 045101 (2006)
30. Ciliberto, S., Imparato, A., Naert, A., Tanase, M.: Heat flux and entropy produced by thermal fluctuations. Phys. Rev. Lett. **110**, 180601 (2013)
31. Abdoli, I., Wittmann, R., Brader, J.M., Sommer, J.-U., Löwen, H., Sharma, A.: Tunable brownian magneto heat pump. Sci. Rep. **12**(1), 13405 (2022)
32. Nyquist, H.: Thermal agitation of electric charge in conductors. Phys. Rev. **32**, 110–113 (1928)
33. Movilla Miangolarra, O., Taghvaei, A., Chen, Y., Georgiou, T.T.: Thermodynamic engine powered by anisotropic fluctuations. Phys. Rev. Res. **4**, 023218 (2022)

Chapter 4
Energy Harvesting from Anisotropic Temperature Fields

Harvesting energy is a principal characteristic of living organisms. Yet, relevant processes rarely conform to the setting of Carnot's engine alternating contact between heat baths of different temperatures. Instead, fluctuations and anisotropic chemical concentrations in conjunction with varying electrochemical potentials seem to provide the universal source of cellular energy [1]. These anisotropies induce non-equilibrium steady states (NESS), which are, in turn, the hallmark of life [2, 3].

In the following chapter, we will study work extraction from far-from-equilibrium transitions that are fueled by anisotropic thermal excitation. Specifically, we will first study a minimal thermodynamic engine built around the concept of the Brownian gyrator [4], in which the characteristic non-equilibrium steady-state circulating current acts as the source of power. This will serve us as a stepping stone in order to introduce a general theory revolving around work extraction that hinges upon the geometry of thermodynamic cycles.

4.1 Harvesting Energy from the Brownian Gyrator

The steady-state characteristics of the Brownian gyrator, such as its non-equilibrium steady-state curl-carrying probability current, the heat flux induced or the torque generated, have been the focus of extensive work [4–10]. In the present section, we take the next natural step and consider energetics of a cyclic operation for such a two-dimensional system. We explore the coupling of its natural gyrating motion with periodic actuation for the purpose of extracting work from the anisotropy of the temperature field [11, 12]. Our main focus is on optimal protocols stemming form conservative actuation [12].

To this end, we extend concepts of thermodynamic geometry [13–15] to regimes far from equilibrium. Specifically, we show that path-lengths traversed in the

O. Movilla Miangolarra, *Stochastic Thermodynamic Treatment of Thermal Anisotropy*, Springer Theses, https://doi.org/10.1007/978-3-031-68066-3_4

manifold of thermodynamic states, measured in a suitable Riemannian metric, represent dissipative losses, while area integrals of a work-density quantify work being extracted. Thus, the maximal amount of work that can be extracted relates to an isoperimetric problem, trading off area against length of an encircling path. Moreover, we derive an isoperimetric inequality that provides a universal bound on the efficiency of all cyclically operating protocols, and a bound on how fast a closed path can be traversed before it becomes impossible to extract positive work. In this way, we quantify trade-offs between efficiency and power that can be extracted.

4.1.1 Model and Analysis

Let us consider a two-dimensional overdamped Brownian particle in an anisotropic heat bath, subject to a time-varying quadratic potential

$$U(t,x) = \frac{1}{2}x'K(t)x,$$

where $x \in \mathbb{R}^2$ and $K(t)$ a symmetric 2×2 matrix, and a linear non-conservative force f which without loss of generality[1] is assumed to be divergence-free with respect to ρ ($\nabla \cdot (f\rho) = 0$), i.e. it is given by

$$f(t, X_t) = \Omega\Sigma_{ss}^{-1}X_t,$$

for Ω a skew-symmetric matrix (possibly time-varying). Thus, the system obeys the Langevin dynamics

$$dX_t = \frac{1}{\gamma}\big(f(t,X_t) - \nabla U(t,X_t)\big)dt + \sqrt{\frac{2k_BT}{\gamma}}dB_t, \tag{4.1}$$

where $\{B_t\}$ is an 2-dimensional Brownian motion, while

$$T = \begin{bmatrix} T_1 & 0 \\ 0 & T_2 \end{bmatrix}.$$

Here, we assume $T_1 > T_2$ and define $\Delta T := T_1 - T_2 > 0$.

If the initial state is Gaussian, $\mathcal{N}(0, \Sigma_0)$ (i.e., with mean 0 and covariance Σ_0), then it remains Gaussian. Its mean remains 0 while the covariance $\Sigma(t)$ satisfies the Lyapunov equation (see Sect. 3.4 for the derivation)

[1] It is without loss of generality since we are interested in the work output over a cycle. Over a cycle, the work output is equal to the heat, which is proportional to the difference $\nabla U - f$. Hence, we can absorb any gradient part of f into ∇U.

$$\gamma \dot{\Sigma}(t) = -K(t)\Sigma(t) - \Sigma(t)K(t) + 2k_B T. \tag{4.2}$$

Since the non-conservative force is divergence-free with respect to ρ, it plays no role in the dynamics of the system (this is due to Ω being skew-symmetric).

The system exchanges energy with the external actuation through work done by changes in the potential or by the presence of a non-conservative force, and through heat transfer with the two thermal baths. The total energy of the system is $\mathcal{E} = \int U\rho\, dx$, while the rate of work is given by two components, one due to changes in the potential and one due to the non-conservative force, i.e.

$$đ\mathcal{W} = \int \frac{\partial U}{\partial t}\rho\, dxdt + \int \langle f, J\rangle dxdt. \tag{4.3}$$

On the other hand, the heat uptake from the thermal baths is

$$đ\mathcal{Q} = đ\mathcal{Q}_1 + đ\mathcal{Q}_2 = \int (\partial_1 U - f_1)J_1 dxdt + \int (\partial_2 U - f_2)J_2 dxdt. \tag{4.4}$$

Our goal in this section is to maximize, over a cycle, the work extraction from the anisotropy in these coupled degrees of freedom. In general, one can do so through both non-conservative actuation and potential forcing. In the following, we will briefly consider extracting work from non-conservative actuation at steady-state. This will turn out to be fairly simple and of independent interest due to its connection to the standard problem in circuit theory of impedance matching [11]. Then, we will turn to the problem of maximizing power output through periodic variation of a potential forcing, which will constitute our main focus for being both physically more meaningful (it opens up the door for autonomous functioning [16]; see Chap. 6) and also experimentally more accessible.

4.1.2 Non-Conservative Forcing at Steady-State

Let us picture the steady-state of the Brownian gyrator (the system described by (4.1) subject to a fixed quadratic potential with $f = 0$) as a vortex of swirling particles around the origin, which originates from the non-zero probability current at steady-state. Now imagine yourself sailing around the origin, propelled by the wind of particles, going in circles at some velocity slower than the particles, yet non-zero, slowed down by external forces applied on the sailboat (e.g. drag). This motion would extract, from the vortex of particles, an amount of work proportional to the forces applied on the sailboat times the displaced distance.

Indeed, one can use non-conservative actuation to mimic this behavior. Let $K(t) = K_c$ be a constant, positive definite matrix. Then, the system reaches a steady state with covariance Σ_{ss}, that together with K_c satisfies

$$\Sigma_{ss} K_c + K_c \Sigma_{ss} = 2k_B T. \tag{4.5}$$

Let us consider having control over a non-conservative force $f = \Omega \Sigma_{ss}^{-1} X_t$, which represents the sailboat, that due to its skew-symmetry, does not alter the steady state of our system. Then, the mean velocity of the particles is given by

$$v(x) = \frac{1}{\gamma}\left(-K_c x + \Omega \Sigma_{ss}^{-1} x + k_B T \Sigma_{ss}^{-1} x\right), \tag{4.6}$$

and the work rate can be expressed as

$$\begin{aligned} đ\mathcal{W} = \int \langle f, v\rho \rangle dx dt &= \frac{1}{\gamma}\mathbb{E}\Big[X_t' \Sigma_{ss}^{-1} \Omega' \left(-K_c + \Omega \Sigma_{ss}^{-1} + k_B T \Sigma_{ss}^{-1}\right) X_t\Big] dt \\ &= \frac{1}{\gamma} \operatorname{Tr}\left[\Omega \Sigma_{ss}^{-1} \Omega' + k_B \Omega \Sigma_{ss}^{-1} T\right] dt, \end{aligned}$$

where we have used the fact that the trace of the product of a symmetric and skew-symmetric matrix is zero.

Therefore, maximizing steady-state work output is equivalent to the static problem[2]

$$\min \operatorname{Tr}[\Omega \Sigma_{ss}^{-1} \Omega' + k_B \Omega \Sigma_{ss}^{-1} T],$$

where the minimum is taken over skew-symmetric Ω's. The first-order necessary condition for optimality is found by computing the first-order variation of the cost and setting it to zero, this is,

$$\operatorname{Tr}[\Delta_\Omega M] = 0, \quad \text{with } M = 2\Omega \Sigma_{ss}^{-1} + k_B T \Sigma_{ss}^{-1}, \tag{4.7}$$

where Δ_Ω is any skew-symmetric matrix. Therefore, the first-order necessary condition for optimality (4.7) is equivalent to imposing M to be symmetric, due to orthogonality. It follows that an optimal choice Ω^* must satisfy

$$\Sigma_{ss}^{-1} \Omega^* + \Omega^* \Sigma_{ss}^{-1} = \frac{k_B}{2}\left(\Sigma_{ss}^{-1} T - T \Sigma_{ss}^{-1}\right). \tag{4.8}$$

This equation has a unique solution which is skew-symmetric and can be written as

$$\begin{aligned} \Omega^* &= \frac{k_B}{2} L_{\Sigma_{ss}^{-1}}(\Sigma_{ss}^{-1} T - T \Sigma_{ss}^{-1}) \\ &= \frac{k_B}{2} \int_0^\infty e^{-\Sigma_{ss}^{-1}\tau}\left(\Sigma_{ss}^{-1} T - T \Sigma_{ss}^{-1}\right) e^{-\Sigma_{ss}^{-1}\tau} d\tau. \end{aligned}$$

[2] The problem is independent of time and, therefore, the optimal Ω will also be time independent.

Note that when Σ_{ss} and T commute, no work can be extracted (as one would expect), since in this case $\Omega^* = 0$.

It is important to highlight that the divergence-free part (the part that is responsible of the vortex-like motion of the particles) of the mean velocity (4.6) comes uniquely from the skew-symmetric part (with respect to ρ) of $\Omega\Sigma_{ss}^{-1} + k_B T \Sigma_{ss}^{-1}$. Therefore, we can interpret the condition of M being symmetric as the fact that the optimal f halves the divergence-free part of the original v (3.17) (the one without f). As a result, the problem of maximizing work output from the Brownian gyrator at steady-state is essentially an impedance matching problem [11].

The maximum work output at steady-state can be computed as follows,

$$\begin{aligned}
đ\mathcal{W}_{\text{out}}^* &= -\frac{1}{\gamma}\,\text{Tr}[\Omega^*\Sigma_{ss}^{-1}\Omega^{*\prime} + k_B\Omega^*\Sigma_{ss}^{-1}T]dt \\
&= \frac{1}{2\gamma}\,\text{Tr}[\Omega^*(\Omega^*\Sigma_{ss}^{-1} + \Sigma_{ss}^{-1}\Omega^*) - k_B\Omega^*(\Sigma_{ss}^{-1}T - T\Sigma_{ss}^{-1})]dt \\
&= \frac{k_B}{4\gamma}\,\text{Tr}[(T\Sigma_{ss}^{-1} - \Sigma_{ss}^{-1}T)\Omega^*]dt \\
&= \frac{k_B}{2\gamma}\,\text{Tr}[T\Sigma_{ss}^{-1}\Omega^*]dt.
\end{aligned}$$

Here, for the third equality we have used (4.8). It is seen that the maximal work is proportional to the anti-symmetric part of $T\Sigma_{ss}^{-1}$, which constitutes the source of energy. Note that this expression can be written explicitly in terms of the parameters of our system: first solve for the steady state covariance Σ_{ss} in (4.5) (which is fairly simple for a two-dimensional problem; see (4.11) for a similar computation), then use it to find Ω^* and finally to write $\mathcal{W}_{\text{out}}^* = \frac{k_B}{2\gamma}\int_0^{t_f} \text{Tr}[T\Sigma_{ss}^{-1}\Omega^*]dt$.

Earlier treatments of this problem can be found in [17–19]. In fact, in [19] such an engine is experimentally realized by suspending a particle in a viscous medium through a feedback optical trap. Optical traps typically induce potential forces, whereas in this setting control over non-conservative forcing is essential. Only recently have we been able to generate these non-conservative forces through feedback optical traps, which allow applying any arbitrary force on the particle by detecting its location to decide what kind of potential would give the necessary force at that particular location [20, Chapter 3], [21]. Unfortunately, this is both energy consuming (measurements, computing, feedback loop) and artificial: one uses the information in the feedback to "pretend" to be applying a non-conservative force instead of a conservative one (potential). Non-conservative forcing can be similarly realized in the electric circuit setting of Fig. 3.1a [11]. However, this can only be obtained through active components (components that require energy input [11, 22]). Indeed, this lack of accounting for the energy input required is intrinsic to the nature of non-conservative forcing.

Thus, in the next section we shift our focus to optimally extracting work through the periodic variation of a potential force, which as we will see in Chap. 6, will allow us to extract energy autonomously.

4.1.3 Conservative Forcing

4.1.3.1 Energetics

Let $f = 0$ and $K(t)$ be our time-varying periodic control. The energy of the system can be written in terms of the state covariance and control as

$$\mathcal{E} = \iint U\rho dxdy = \frac{1}{2}\operatorname{Tr}[K(t)\Sigma(t)].$$

The rates of work input (4.3) and total heat input (4.4) in the absence of a non-conservative force simplify to

$$đ\mathcal{W} = \frac{1}{2}\operatorname{Tr}[\dot{K}(t)\Sigma(t)]dt \quad \text{and} \quad đ\mathcal{Q} = \frac{1}{2}\operatorname{Tr}[K(t)\dot{\Sigma}(t)]dt.$$

Our goal in this section is to design $K(t)$ so as to extract the maximum amount of work possible by steering the covariance matrix $\Sigma(t)$ along a closed trajectory with $\{\Sigma(t); t \in [0, t_f], \Sigma(0) = \Sigma(t_f)\}$ in a cyclic manner.

To simplify our analysis we consider $\dot{\Sigma}(t)$ as our design parameter, instead of $K(t)$. We can do so since the unique $K(t)$ that satisfies (4.2) is obtained in terms of $(\dot{\Sigma}(t), \Sigma(t))$ as

$$K(t) = L_{\Sigma(t)}[2k_B T - \gamma\dot{\Sigma}(t)].$$

Then, the heat rate, also expressed in terms of $(\Sigma(t), \dot{\Sigma}(t))$, is

$$đ\mathcal{Q} = k_B \operatorname{Tr}\left[L_{\Sigma(t)}[T]\dot{\Sigma}(t)\right]dt - \frac{\gamma}{2}\operatorname{Tr}\left[L_{\Sigma(t)}[\dot{\Sigma}(t)]\dot{\Sigma}(t)\right]dt.$$

Integrating over $[0, t_f]$ we obtain that $\mathcal{Q} = \mathcal{Q}_{\text{qs}} + \mathcal{Q}_{\text{diss}}$, where

$$\mathcal{Q}_{\text{qs}} = k_B \int_0^{t_f} \operatorname{Tr}\left[L_{\Sigma(t)}[T]\dot{\Sigma}(t)\right]dt, \tag{4.9a}$$

$$\mathcal{Q}_{\text{diss}} = -\frac{\gamma}{2}\int_0^{t_f} \operatorname{Tr}\left[L_{\Sigma(t)}[\dot{\Sigma}(t)]\dot{\Sigma}(t)\right]dt. \tag{4.9b}$$

These are integrals along the curve $\{\Sigma(t) \mid t \in [0, t_f]\}$. Note that the first one is independent of the time parameter, since it can be written as a line integral, this is, as the integral of the vector field $L_{\Sigma(t)}[T]$ along a curve in Σ-space. On the other hand, the second expression is quadratic in the speed $\dot{\Sigma}$ and thus converges to zero as the speed in traversing the path converges to zero. Therefore, the first expression corresponds to the effective heat uptake in the quasi-static limit, while the second corresponds to dissipation. When integrating over a cycle, the work output is

$$\mathcal{W}_{\text{out}} = \mathcal{Q}_{\text{qs}} - \mathcal{W}_{\text{diss}},$$

where $\mathcal{W}_{\text{diss}} = -\mathcal{Q}_{\text{diss}}$. Moreover, we define the efficiency of the cycle as the ratio between the work output and the maximum amount of work that can be extracted over the cycle in the quasi-static regime [15] , i.e.,

$$\eta = \frac{\mathcal{W}_{\text{out}}}{\mathcal{Q}_{\text{qs}}}.$$

Clearly, $\eta \leq 1$, where equality is achieved in the quasi-static regime (as $t_f \to \infty$).

4.1.3.2 A Two-Dimensional Parametrization

From this point on, we restrict the controlled degrees of freedom on the state manifold (Σ-space) to two, by imposing that $\det(\Sigma(t))$ (or equivalently the entropy of the state) be constant. This restriction is natural since, otherwise, optimal cycles trace singular (delta function) distributions, leading to vanishingly small dissipation, that can only be obtained with arbitrarily large controls (forces) [23]. Moreover, one can show (by following the same steps as the ones presented below) that changing the entropy of the state over a cycle results in zero quasi-static work output [23]. Consequently, we consider constant entropy along a cycle and express the 2×2 positive definite covariance (state) in polar coordinates $(r, \theta) \in [0, \infty) \times [0, 2\pi)$ as

$$\Sigma(r, \theta) = R\Big(-\frac{\theta}{2}\Big)\sigma^2(r)R\Big(\frac{\theta}{2}\Big), \tag{4.10}$$

where $R(\cdot)$ and $\sigma^2(\cdot)$ are orthogonal and diagonal matrices, respectively, given by

$$R(\vartheta) = \begin{bmatrix} \cos(\vartheta) & \sin(\vartheta) \\ -\sin(\vartheta) & \cos(\vartheta) \end{bmatrix} \text{ and } \sigma^2(r) = \begin{bmatrix} \ell_c{}^2 e^r & 0 \\ 0 & \ell_c{}^2 e^{-r} \end{bmatrix},$$

where $\ell_c = \sqrt[4]{\det(\Sigma(t))}$ is a (constant) *characteristic length* for the system. Therefore, the rate $\dot{\Sigma}$ can be expressed as the sum of two terms, one accounting for the rotation and the other for the expansion/contraction, that is,

$$\dot{\Sigma} = \frac{1}{2}R'(\sigma^2\hat{\Omega} - \hat{\Omega}\sigma^2)R\dot{\theta} + R'\sigma^2\Xi R\dot{r},$$

where

$$\hat{\Omega} = \begin{bmatrix} 0 & 1 \\ -1 & 0 \end{bmatrix} \quad \text{and} \quad \Xi = \begin{bmatrix} 1 & 0 \\ 0 & -1 \end{bmatrix}.$$

Substituting this expression for $\dot{\Sigma}$ and (4.10) into (4.9a),

$$\begin{aligned} \mathcal{Q}_{\text{qs}} &= \frac{k_B}{2} \int_0^{t_f} \text{Tr}\left[\int_0^{\infty} e^{-R'\sigma^2 R\tau} T e^{-R'\sigma^2 R\tau} d\tau R' ((\sigma^2 \hat{\Omega} - \hat{\Omega}\sigma^2)\dot{\theta} + \sigma^2 \Xi \dot{r}) R \right] dt \\ &= \frac{k_B}{2} \int_0^{t_f} \text{Tr}\left[N(r,\theta) \big((\sigma^2 \hat{\Omega} - \hat{\Omega}\sigma^2)\dot{\theta} + \sigma^2 \Xi \dot{r} \big) \right] dt, \end{aligned} \tag{4.11}$$

where $N(r,\theta)$ is given by

$$\begin{aligned} &N(r,\theta) \\ &= \int_0^{\infty} e^{-\sigma^2 \tau} R T R' e^{-\sigma^2 \tau} d\tau \\ &= \int_0^{\infty} \begin{bmatrix} e^{-2\ell_c^2 e^r \tau} \big(\cos^2(\theta) T_1 + \sin^2(\theta) T_2 \big) & e^{-\ell_c^2 (e^r + e^{-r})\tau} \sin(2\theta) \big(\frac{T_2 - T_1}{2} \big) \\ e^{-\ell_c^2 (e^r + e^{-r})\tau} \sin(2\theta) \big(\frac{T_2 - T_1}{2} \big) & e^{-2\ell_c^2 e^{-r} \tau} \big(\cos^2(\theta) T_2 + \sin^2(\theta) T_1 \big) \end{bmatrix} d\tau \\ &= \frac{1}{2\ell_c^2} \begin{bmatrix} e^{-r} \big(\cos^2(\theta) T_1 + \sin^2(\theta) T_2 \big) & \frac{2}{e^r + e^{-r}} \sin(2\theta) \big(\frac{T_2 - T_1}{2} \big) \\ \frac{2}{e^r + e^{-r}} \sin(2\theta) \big(\frac{T_2 - T_1}{2} \big) & e^r \big(\cos^2(\theta) T_2 + \sin^2(\theta) T_1 \big) \end{bmatrix}. \end{aligned}$$

Taking the trace of the product of the remaining matrices in (4.11), we obtain that the quasi-static heat can be expressed in polar coordinates as

$$\mathcal{Q}_{\text{qs}} = \frac{k_B \Delta T}{2} \int_0^{t_f} \big(\cos(\theta) \dot{r} - \tanh(r) \sin(\theta) \dot{\theta} \big) dt. \tag{4.12}$$

Following similar computations, dissipation can be written as

$$\mathcal{W}_{\text{diss}} = \frac{\gamma \ell_c^2}{2} \int_0^{t_f} \left(\cosh(r) \dot{r}^2 + \sinh(r) \tanh(r) \dot{\theta}^2 \right) dt. \tag{4.13}$$

4.1.3.3 Geometric Interpretations

We now consider the integrals ((4.12)–(4.13)) over a cycle that encircles a domain $\mathcal{D}$, that is, over the boundary $\partial\mathcal{D}$ of $\mathcal{D}$. Using Stokes' theorem, $\mathcal{Q}_{\text{qs}}$ can be expressed as an area integral over $\mathcal{D}$,

$$\mathcal{Q}_{\text{qs}} = \frac{k_B \Delta T}{2} \oint_{\partial\mathcal{D}} (\cos(\theta) dr - \tanh(r) \sin(\theta) d\theta)$$

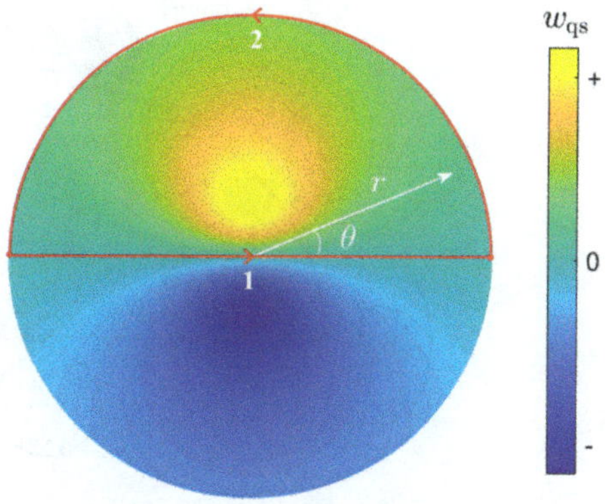

Fig. 4.1 Work-density (4.15) with values color-coded, expressed in state-coordinates (r, θ) in (4.10). Area integrals over closed cycles represent quasi-static work. The red cycle encompasses the region of positive work-density within a given radius. Image extracted from [12]

$$= \pm \frac{k_B \Delta T}{2} \iint_{\mathcal{D}} \frac{\tanh^2(r)}{r} \sin(\theta) r d\theta dr, \tag{4.14}$$

where the sign is positive if the direction in traversing the cycle is counterclockwise (CCW), and negative otherwise. Thus,

$$w_{qs}(r, \theta) = \frac{k_B \Delta T}{2} \frac{\tanh^2(r)}{r} \sin(\theta) \tag{4.15}$$

represents a quasi-static work-density, which is depicted in Fig. 4.1, and is positive on the upper half plane and negative on the lower. Any CCW cycle encircling a domain in the upper half plane results in positive work output. Likewise, a CW cycle in the lower half plane results in positive work output as well. The opposite is true when the flow is reversed. Below we always consider CCW-cycles.

The dissipation (4.13) can be written as the (action) integral

$$\mathcal{W}_{\text{diss}} = \frac{\gamma \ell_c^2}{2} \int_0^{t_f} \|\dot{\alpha}(t)\|_{g_W}^2 dt,$$

where $\{\alpha(t) = (r(t), \theta(t)) \mid t \in [0, t_f]\}$ traces $\partial\mathcal{D}$, and $\|\dot{\alpha}\|_{g_W}^2 := \dot{\alpha}' g_W \dot{\alpha}$ is the square norm of the velocity with respect to the Riemannian metric

$$g_W = \begin{bmatrix} \cosh(r) & 0 \\ 0 & \sinh(r)\tanh(r) \end{bmatrix}.$$

By the Cauchy-Schwartz inequality, one obtains

$$\mathcal{W}_{\text{diss}} \geq \frac{\gamma \ell_c^2}{2t_f} \left(\int_0^{t_f} \|\dot{\alpha}(t)\|_{g_W} dt \right)^2, \tag{4.16}$$

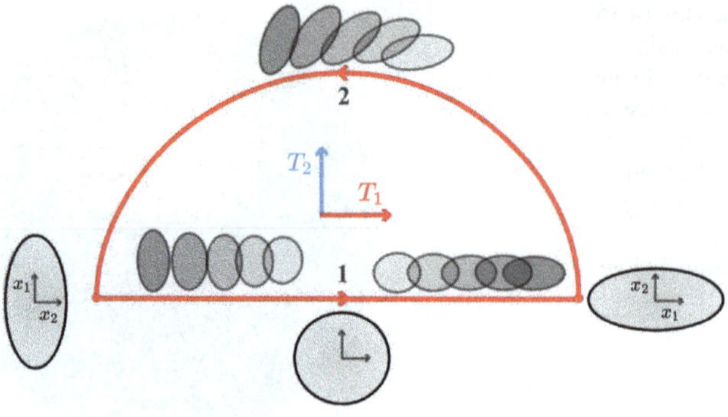

Fig. 4.2 Cyclic protocol for the semicircle path of radius $r_{\max}$ in Figure 4.1. The two phases represent: (1) expansion of the $\Sigma(t)$-ellipsoid along the x_1-axis with simultaneous compression along the x_2-axis, and (2) rotation to bring $\Sigma(t)$ to the starting value. The area of the ellipsoid remains constant during the cycle. When $T_1 > T_2$, work is extracted during phase 1 and added during phase 2. Image extracted from [12]

where equality holds when $\|\dot{\alpha}(t)\|_{g_W}$ remains constant along a curve. The integral in parentheses is the length of the closed curve $\{\alpha(t); t \in [0, t_f]\}$ in the metric g_W, which is nothing but its Wasserstein-2 length (c.f. (1.3)). Therefore, the minimum dissipation is equal to the square of the Wasserstein-2 length of the curve (modulo a factor $\gamma\ell_c^2/2t_f$) achieved when $\|\dot{\alpha}(t)\|_{g_W}$ is constant. From here on, we denote by $\mathcal{M}$ the Riemannian manifold of thermodynamic states equipped with the metric g_W.

The above results are exemplified in Figs. 4.2 and 4.3. Specifically, Fig. 4.2 displays $\Sigma(t)$-ellipsoids, relative to the principal axes of T, for the semicircle cycle (red) of Fig. 4.1. Then, Fig. 4.3 displays efficiency and work output for the same cycle, as a function of the radius of the semicircle, with the period of cycle fixed at $t_f = 2 \cdot 10^{-3}$. Here, work is computed by subtracting the dissipation for constant velocity (RHS of (4.16)) from the quasi-static work (4.14). Moreover, in Fig. 4.3 we observe that an optimal value for $r_{\max}$ balances the two terms, the increase in area against increase in the perimeter, so as to maximize work output. This observation exposes an inherent isoperimetric problem that we discuss next.

4.1.3.4 Isoperimetric Problem

Define the (weighted) area of $\mathcal{D}$ and its perimeter by

$$\mathcal{A}_w = \iint_{\mathcal{D}} w(r, \theta)\sqrt{\det(g_W)}d\theta dr, \quad \ell = \oint_{\partial\mathcal{D}} ds,$$

respectively, where $ds = \|\dot{\alpha}(t)\|_{g_W} dt$, and

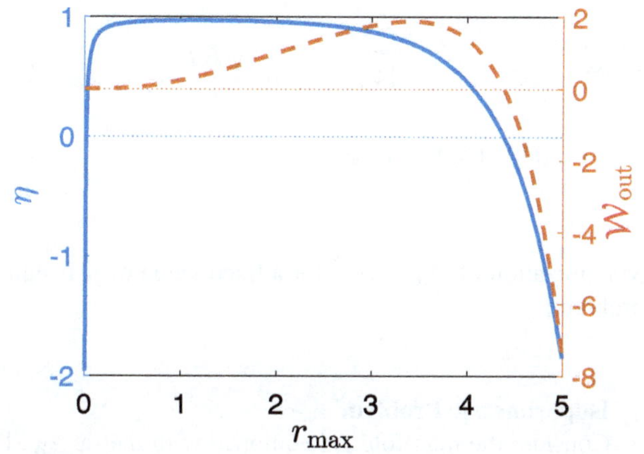

Fig. 4.3 Efficiency and work output for the cycle depicted in Fig. 4.2 as a function of the radius of the semicircle $r_{\max}$. Image from [12]

$$w(r, \theta) = \frac{\sin(\theta)\sinh(r)}{\cosh^2(r)}$$

is a work-density relative to the Riemannian canonical 2-form ("unit area") $\sqrt{\det(g_W)}d\theta dr$. The area and the perimeter characterize the quasi-static heat $\mathcal{Q}_{\text{qs}}$ and dissipation $\mathcal{W}_{\text{diss}}$, as

$$\mathcal{Q}_{\text{qs}} = \frac{k_B \Delta T}{2}\mathcal{A}_w, \quad \mathcal{W}_{\text{diss}} = \frac{\gamma \ell_c^2}{2t_f}\ell^2,$$

and these determine the work output $\mathcal{W}_{\text{out}}$ and efficiency η, as

$$\mathcal{W}_{\text{out}} = \frac{k_B \Delta T}{2}(\mathcal{A}_w - \mu\ell^2) \quad \text{and} \quad \eta = 1 - \mu\frac{\ell^2}{\mathcal{A}_w}, \tag{4.17}$$

where $\mu = \frac{t_c}{t_f}$ is a dimensionless constant, with $t_c = \frac{\gamma \ell_c{}^2}{k_B \Delta T}$ the *characteristic time* that a Brownian motion with intensity $\sqrt{\gamma^{-1}k_B \Delta T}$ needs to traverse a distance ℓ_c on average.

We now consider maximizing work output over cycles on the manifold of thermodynamic states, i.e.,

Problem of Maximizing Work Output
Consider the two-dimensional Langevin system (4.1) under a quadratic potential. Determine the closed cycle, parametrized with constant velocity, that maximizes

(continued)

$$\mathcal{W}_{\text{out}}^*(\mu) := \frac{k_B \Delta T}{2} \max_{\mathcal{D}} \{\mathcal{A}_w - \mu \ell^2\}, \tag{4.18}$$

for different values of μ.

Maximization of $\mathcal{A}_w - \mu \ell^2$ for a fixed value of μ is equivalent to an isoperimetric problem:

Isoperimetric Problem
Consider the manifold $\mathcal{M}$ equipped with metric g_W. Determine the curve on $\mathcal{M}$ of fixed length ℓ that maximizes the enclosed (weighted) area, this is,

$$\mathcal{A}_w^*(\ell) := \max_{\mathcal{D}} \{\mathcal{A}_w \mid \ell \text{ is specified }\}. \tag{4.19}$$

The relationship between both problems becomes clear if we note that μ in (4.18) can be seen as a Lagrange multiplier for (4.19), where the term $\mu \ell^2$ acts as a penalty on the length. Consequently, the larger the parameter μ in (4.18), the shorter the associated optimal length. Conversely, the shorter the fixed length ℓ in (4.19), the larger the associated penalty μ. Therefore, it is intuitively clear that for any fixed parameter μ, there exists a length ℓ, such that (4.18) and (4.19) are equivalent, and vice-versa.

We obtain a first-order condition that characterizes optimal cycles (for either problem) through variational analysis. To this end, we parametrize the closed curve $\alpha(\cdot)$ tracing $\partial \mathcal{D}$ by the arclength s. We let ds and du denote the differentials along the curve and normal to the curve respectively, so that the corresponding local coordinates form an orthonormal system. Under a perturbation $\alpha(s) \to \alpha(s) + \psi(s)\hat{n}(s)du$, where $\hat{n}(s)$ is the (outward) normal unit vector at s and $\psi(\cdot)$ is an arbitrary scalar function, the perimeter is perturbed to $\int_{s=0}^{\ell}(1 + \kappa(s)\psi(s)du)ds$, where $\kappa(\cdot)$ denotes the geodesic curvature [24]. Thus, the variation of ℓ^2 is $\delta \ell^2 = 2\ell \int_{s=0}^{\ell} \kappa(s)\psi(s)dsdu$. On the other hand, as the domain $\mathcal{D}$ is enlarged to $\mathcal{D} \cup \delta \mathcal{D}$,

$$\delta \mathcal{A}_w = \iint_{\delta \mathcal{D}} w(r, \theta)\sqrt{\det(g_W)}d\theta dr = \int_{s=0}^{\ell} w(r(s), \theta(s))\psi(s)dsdu.$$

Hence, the first-order optimally condition $\delta \mathcal{A}_w - \mu \delta \ell^2 = 0$ gives that the ratio of the geodesic curvature κ over the density w must be constant and equal to $1/(2\ell\mu)$ at each point of the curve that traces $\partial \mathcal{D}$, i.e.

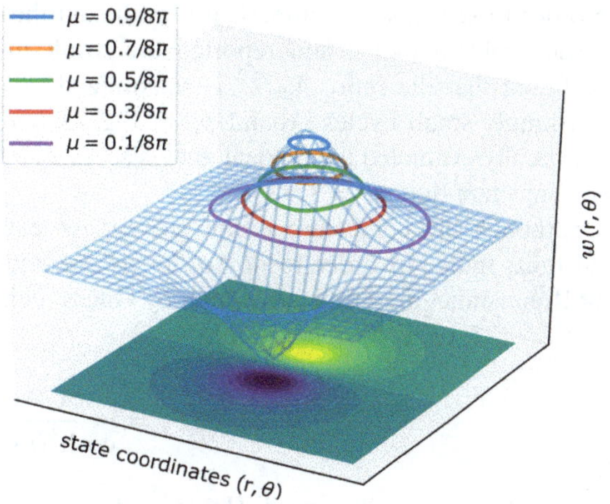

Fig. 4.4 Optimal cycles, in polar coordinates (r, θ), that maximize work output for different values of μ. The cycles are drawn on the w-density surface and solve an isoperimetric problem. Image extracted from [12]

$$\frac{\kappa(r,\theta)}{w(r,\theta)} = \frac{1}{2\ell\mu}.$$

Figure 4.4 displays several such optimal curves that have been obtained numerically using the first-order optimality condition. It is observed that as μ becomes small, and thus, the corresponding penalty on the length decreases, the area that the optimal cycle encircles increases. On the other hand, as μ becomes large, the optimal cycle shrinks to the point $p_0 = (r_0, \theta_0) = (\sinh^{-1}(1), \frac{\pi}{2})$, beyond which (i.e., for larger μ) it is impossible to extract positive work. The point p_0 is where w achieves its maximum.

4.1.3.5 Isoperimetric Inequality

The impossibility of extracting positive work for large values of μ points to an isoperimetric inequality, this is, a bound on the ratio between the area and the square of the perimeter of any curve:

Isoperimetric Inequality
Consider the manifold $\mathcal{M}$ equipped with metric g_W. For all closed curves, the ratio between area and perimeter-squared is bounded, with

$$\mu^* := \sup_{\mathcal{D}} \left\{ \frac{\mathcal{A}_w}{\ell^2} \right\} < \infty, \tag{4.20}$$

being the isoperimetric constant.

In order to see this, we numerically evaluated the function $\mathcal{A}_w^*(\ell)$ in the isoperimetric problem (4.19) and reported the result in Fig. 4.5. It can be seen from the figure that the ratio $\mathcal{A}_w/\ell^2$ is maximized as $\ell \to 0$, which corresponds to vanishingly small cycles around p_0 in Fig. 4.4. For such cycles, the ratio can be analytically evaluated using local analysis, $\mathcal{A}_w/\ell^2 \simeq w(p_0)/(4\pi) = 1/(8\pi)$. Thus we conjecture that $\mu^* = 1/(8\pi)$.

Although the conjecture is not proven, we can establish that $\mu^* \leq 1/(4\pi)$, and thus, finite. In order to do so, let us consider the Gaussian curvature G of our Riemannian manifold $\mathcal{M}$, which by Gauss' celebrated theorema egregium [24, page 23],[25, page 400], is given by

$$G(r, \theta) = \frac{1}{\cosh^3(r)}.$$

Thus, $\mathcal{M}$ is positively curved. However, the curvature decreases radially to 0. For such manifolds, where g_W is rotationally symmetric and the Gaussian curvature is radially decreasing, the following isoperimetric inequality holds [24, Page 113],

$$\ell^2 \geq 4\pi\mathcal{A} - 2\int_0^{\mathcal{A}} \bar{G}(\tau)d\tau, \tag{4.21}$$

where

$$\mathcal{A} = \iint_{\mathcal{D}} \sqrt{\det(g_W)}drd\theta,$$

is the (unweighted) area of $\mathcal{D}$ with respect to the canonical 2-form of $\mathcal{M}$, and $\bar{G}(\tau)$ is the area integral of the Gaussian curvature over a circle centered at the origin with area τ. This circle has radius $r(\tau) = \cosh^{-1}(1 + \frac{\tau}{2\pi})$. Therefore,

$$\bar{G}(\tau) = \int_0^{r(\tau)} \int_0^{2\pi} \frac{\sinh(r)}{\cosh(r)^3} d\theta dr = \pi \left(1 - \frac{1}{(\frac{\tau}{2\pi} + 1)^2}\right).$$

Using this result in (4.21),

$$\ell^2 \geq 2\pi\mathcal{A} + \frac{4\pi^2\mathcal{A}}{\mathcal{A} + 2\pi} \geq 2\pi\mathcal{A}.$$

Since $\mathcal{A}_w \leq \max_{(r,\theta)\in\mathcal{D}} w(r, \theta) \times \mathcal{A} = \frac{1}{2}\mathcal{A}$, we conclude that

$$\mu^* \leq \frac{1}{4\pi}.$$

This bound is not tight due to the fact that w is not rotationally symmetric (as opposed to the curvature) and achieves its maximum at $(r_0, \theta_0) = (\sinh^{-1}(1), \frac{\pi}{2})$.

The isoperimetric inequality (4.20) has two important implications. First, for $\mu \geq \mu^*$ (equivalently, $t_f \leq t_c/\mu^*$), it is impossible to extract positive work. Thus, $\frac{t_c}{\mu^*}$ constitutes a threshold for the period of work producing cycles. Second, the efficiency is bounded by

$$\eta \leq 1 - \frac{\mu}{\mu^*} = 1 - \frac{1}{\mu^*}\frac{t_c}{t_f}.$$

The bound depends on physical parameters and the period, and turns negative when positive work output is not possible.

Moreover, the shape of $\mathcal{A}_w^*(\ell)$ helps answer a variety of questions on optimizing protocols. Specifically, the maximal work output $\mathcal{W}_{\text{out}}^*(\mu)$ in (4.18) corresponds to the maximal vertical distance between $\mathcal{A}_w^*(\ell)$ and the line $\mu\ell^2$, which takes place where $d\mathcal{A}_w^*(\ell)/d\ell^2 = \mu$. Also, it allows computing the maximal work for a given efficiency η. Operating points with efficiency η provide work $\mathcal{W}_{\text{out}} = \eta\mathcal{A}_w$ and lie on the line $\mathcal{A}_w = \frac{\mu}{1-\eta}\ell^2$ shown (dash-dotted) in Fig. 4.5. Therefore, the intersection of this line with the (blue) curve $\mathcal{A}_w^*(\ell)$ in Fig. 4.5 gives the sought optimal operating point for a given efficiency.

In the above, we tacitly assumed that the curve $\mathcal{A}_w^*(\ell)$ intersects any line $\mu\ell^2$, for $\mu < \mu^*$, and that it eventually stays below the line, in that $\lim_{\ell\to\infty} \mathcal{A}_w/\ell^2 = 0$. We show that this is indeed true by proving the bound

$$\mathcal{A}_w - \mu\ell^2 \leq \frac{1}{4\mu}, \tag{4.22}$$

for all $\mu > 0$. This bound is established through a completion of squares

$$\begin{aligned}\mathcal{W}_{\text{out}} = \frac{k_B\Delta T}{2}\int_0^{t_f} &\{\cos(\theta)\dot{r} - \tanh(r)\sin(\theta)\dot{\theta} - t_c\cosh(r)\dot{r}^2 \\ &- t_c\sinh(r)\tanh(r)\dot{\theta}^2\}dt\end{aligned}$$

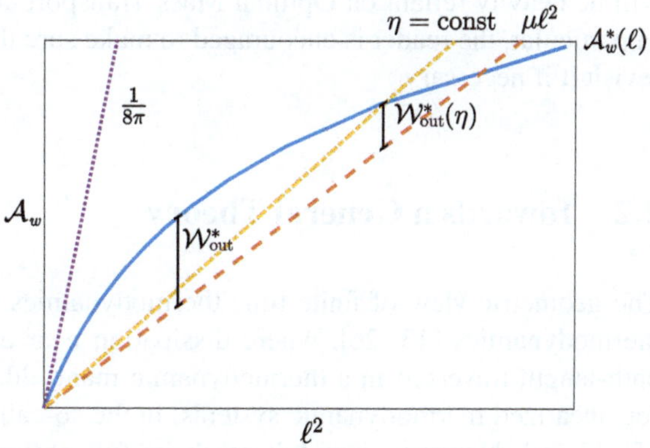

Fig. 4.5 Solution $\mathcal{A}_w^*(\ell)$ to the isoperimetric problem (4.19), shown with solid blue curve as a function of the fixed length-squared ℓ^2. Image extracted from [12]

$$
\begin{aligned}
&= \frac{k_B \Delta T}{2} \int_0^{t_f} \Bigg\{ -\left(\sqrt{t_c \cosh(r)}\, \dot{r} - \frac{\cos(\theta)}{2\sqrt{t_c \cosh(r)}} \right)^2 \\
&\quad - \left(\sqrt{t_c \sinh(r) \tanh(r)}\, \dot{\theta} + \frac{\sin(\theta)}{2\sqrt{t_c \cosh(r)}} \right)^2 + \frac{1}{4t_c \cosh(r)} \Bigg\} dt \\
&\le \frac{k_B \Delta T}{8t_c} \int_0^{t_f} \frac{1}{\cosh(r)} dt \le \frac{k_B \Delta T}{8\mu},
\end{aligned}
$$

where in the first step we have completed the squares, and for the last inequality we used the fact that $\cosh(r) \ge 1$. Taking $\mu = \frac{1}{2\ell}$ in (4.22), we have that $\mathcal{A}_w \le \ell$, concluding that $\lim_{\ell \to \infty} \mathcal{A}_w / \ell^2 = 0$. Another consequence of (4.22) is that the power output is bounded as well, since

$$
\mathcal{P} = \frac{\mathcal{W}_{\text{out}}}{t_f} = \frac{k_B \Delta T}{2t_f} (\mathcal{A}_w - \mu \ell^2) \le \frac{k_B \Delta T}{8t_c}.
$$

It is important to note that this bound on power is independent of the chosen cycle or its period t_f, and only depends on the ratio between the energy $k_B \Delta T$ and the characteristic time t_c. We also remark that, whereas the work extracted is monotone in the period t_f (the larger t_f the larger the work extracted), the relationship between t_f and power is more involved. In particular, one can optimize power over t_f to obtain [23]

$$
\mathcal{P}^* = \frac{k_B \Delta T}{8t_c} \frac{\mathcal{A}_w^2}{\ell^2},
$$

where the optimal power is achieved for $t_f = 2t_c \frac{\ell^2}{\mathcal{A}_w}$.

So far we have studied optimal protocols that extract maximum work from a linear two-dimensional system. We are now ready to generalize these results to higher dimensions. In doing so, we will develop a geometric theory of stochastic thermodynamics that accounts for transitions away from equilibrium. This theory will be heavily reliant on Optimal Mass Transport concepts introduced in Chap. 1. In particular, the reader is encouraged to make sure they understood Sect. 1.4 and to revisit it if necessary.

4.2 Towards a General Theory

The geometric view of finite time thermodynamics has its origins in macroscopic thermodynamics [13, 26], where dissipation near equilibrium is understood as a path-length traversed in a thermodynamic manifold. It received renewed attention for linearized thermodynamic systems, in the so-called linear response regime [14, 15, 27–29]. However, it didn't reach its full potential until the relation between

Stochastic Control, Thermodynamics, and Optimal Mass Transport was drawn [30–33]. As we have seen, this relation enabled the understanding of entropy production as the square of a Riemannian length, the Wasserstein-2 length.

Herein, we push this geometric view a step further. Using the previous section as a guideline, we present a general analysis of the cyclic operation of a heat engine powered by anisotropy in its thermal environment [34]. Specifically, we show that, for an n-dimensional overdamped system driven by a potential force, dissipation and quasi-static work can be expressed as path integrals on the manifold of thermodynamic states. Our interest is in maximizing work output while driving the overdamped system over a cycle by periodic control. When the trajectory of our system lies on a two-dimensional manifold, the quasi-static work extracted over a cycle can be expressed as an area integral. Therefore, the problem of maximizing work output over a cycle amounts to maximizing an area integral subject to a fixed length of the closed thermodynamic path, which (again) constitutes an isoperimetric problem. As a consequence, limits to efficiency are intrinsically related to an isoperimetric inequality.

Analogous isoperimetric results were derived in the linear response regime in [29, 35, 36], and in the context of quantum thermodynamics in [37]. We wish to emphasize that the results presented here apply to classical systems outside the linearized regime.

4.2.1 Model and Analysis

We consider controlled n-dimensional overdamped Langevin dynamics

$$\gamma dX_t = -\big(\nabla U(t, X_t) - f(t, X_t)\big)dt + \sqrt{2\gamma k_B T}dB_t, \tag{4.23}$$

where $X_t \in \mathbb{R}^n$, $T = \mathrm{diag}(T_1, \dots, T_n)$, with T_i's distinct in order to allow work extraction, $\{B_t\}$ is an n dimensional standard Brownian motion, $U(t, x)$ represents a time-varying potential exerting a force $-\partial_{x_i} U(t, x)$ along the i-th degree of freedom, while $f(t, x)$ represents a non-conservative force.

The probability density function of X_t, denoted by $\rho(t, x)$, constitutes the state of the system (thermodynamic ensemble) and satisfies the Fokker-Planck equation, repeated here for convenience,

$$\partial_t \rho + \nabla \cdot (\rho v) = 0, \tag{4.24}$$

where

$$v = -\frac{1}{\gamma}(\nabla U - f + k_B T \nabla \log(\rho)), \tag{4.25}$$

with f such that $\nabla \cdot (f\rho) = 0$, i.e., divergence-free[3] with respect to ρ.

The differential of work into the system due to both changes in the potential and non-conservative forcing is

$$đ\mathcal{W} = \int \frac{\partial U}{\partial t}\rho\, dxdt + \int \langle f, J\rangle dxdt. \tag{4.26}$$

The heat uptake from the respective thermal baths is

$$đ\mathcal{Q}_i = \int (\partial_i U - f_i) J_i dxdt,$$

resulting in the total heat uptake

$$đ\mathcal{Q} = \sum_{i=1}^{n} đ\mathcal{Q}_i = \int \langle \nabla U - f, J\rangle dxdt. \tag{4.27}$$

Our objective is to maximize the work output over a cycle. Once again, we will focus in doing so through a potential function, which will lead to an isoperimetric problem. However, before we dive into this, let us first consider the simpler problem of maximizing work output at steady-state when we have control over non-conservative forces.

4.2.2 Non-Conservative Forcing

Let ρ_{ss} be an arbitrary state satisfying Assumption A1 (Sect. 1.4). We would like to first find a time-independent potential U_c that makes ρ_{ss} a steady-state. The potential must satisfy

$$\nabla \cdot (v\rho_{ss}) = 0, \tag{4.28}$$

with

$$v = -\frac{1}{\gamma}(\nabla U_c - f + k_B T \nabla \log \rho_{ss}).$$

This is equivalent to imposing

[3] This is without loss of generality, as in the previous section, since we are interested in the work output over a cycle. Over a cycle, the work output is equal to the heat, which is proportional to the difference $\nabla U - f$. Hence, we can absorb any gradient part of f into ∇U.

$$\nabla \cdot (\rho_{ss} \nabla U_c) = -\nabla \cdot (\rho_{ss} k_B T \nabla \log(\rho_{ss})),$$

which is nothing but a Poisson equation for U_c and thus has a unique solution, since the Poincaré inequality is satisfied and the square norm of the (weighted) right-hand-side, $\|\frac{\nabla \cdot (T \nabla \rho_{ss})}{\rho_{ss}}\|^2_{\rho_{ss}}$, is bounded by Assumption A1 (see Sect. 1.4). Equivalently, ∇U_c is fixed by the gradient part of $k_B T \nabla \log(\rho_{ss})$, i.e. $\nabla U_c = -\Pi_{\rho_{ss}} k_B T \nabla \log(\rho_{ss})$.

With the potential fixed, the work due to the non-conservative forcing is given by

$$đ\mathcal{W} = \frac{1}{\gamma} \int f'(-\nabla U_c + f - k_B T \nabla \log \rho_{ss}) \rho_{ss} dx dt.$$

The optimal f that maximizes work output must be such that the first-order variation of $đ\mathcal{W}$,

$$\frac{1}{\gamma} \int (\delta_f)'(-\nabla U_c + 2f - k_B T \nabla \log \rho_{ss}) \rho_{ss} dx dt,$$

vanishes for all admissible δ_f. Given that δ_f must be divergence-free with respect to ρ_{ss}, we obtain that, due to orthogonality,

$$-\nabla U_c + 2f - k_B T \nabla \log \rho_{ss}$$

must be of gradient form. This expression is simply $v + f$. Since ρ_{ss} is a steady-state, the gradient part of v must vanish (i.e. must satisfy (4.28)). Given that f is by definition divergence-free, the condition of $v + f$ being of gradient form implies that it must vanish, leading to

$$f^* = \frac{1}{2}(\nabla U_c + k_B T \nabla \log \rho_{ss}).$$

Once again, we can interpret this as the requirement on the optimal f of reducing the divergence-free swirling motion of the particles by half, in a way akin to the standard impedance matching problem [11].

4.2.3 Conservative Forcing

4.2.3.1 Energetics

Let us now go back to our main focus of maximizing work output over potential-driven cycles. Given that work and heat over a cycle are the same up to a sign, consider the total heat uptake which, in the absence of non-conservative forces, can be written as

$$đ\mathcal{Q} = \int \langle \nabla U, v\rho \rangle dxdt = \int \langle \nabla U, \nabla \phi\, \rho \rangle dxdt,$$

where we have used the fact that $\nabla\phi$ is the gradient part of v, $\Pi_\rho v$, which is the only component of the mean velocity affecting the total heat (due to orthogonality). As we saw in the introductory Sect. 1.4, $\nabla\phi$ is in the tangent space of the space of probability distributions $P_2(\mathbb{R}^n)$. In fact, ϕ is given by $\phi = -(U + \varphi)/\gamma$, where φ is the unique solution to $\nabla \cdot (\rho \nabla \varphi) = \nabla \cdot (\rho k_B T \nabla \log(\rho))$.

One can use (4.25) to write the gradient of the potential in terms of the state of the system, i.e., $\nabla U = -k_B T \nabla \log(\rho) - \gamma \nabla \phi - \gamma \chi$, where we have written v as $\nabla\phi + \chi$ with χ divergence-free with respect to ρ. Thus, the total heat uptake becomes

$$đ\mathcal{Q} = -k_B \int (\nabla \log(\rho))' T' \nabla \phi\, \rho\, dxdt - \gamma \int \|\nabla \phi\|^2 \rho\, dxdt,$$

where we have used the fact that $\nabla \cdot (\chi\rho) = 0$. Integrating over the time interval $[0, t_f]$,

$$\mathcal{Q} = -k_B \int_0^{t_f} \langle T \nabla \log(\rho), \nabla \phi \rangle_\rho dt - \gamma \int_0^{t_f} \int \|\nabla \phi\|^2 \rho dxdt,$$

is expressed in terms of ρ and $\nabla\phi$, with $\langle v_1, v_2 \rangle_\rho = \int v_1' v_2 \rho\, dx$ denoting inner product of vector fields.

Note that the integration in the last expression for $\mathcal{Q}$ takes place over a time-indexed path $\{\rho(t, \cdot) \mid t \in [0, t_f]\}$ on the thermodynamic manifold. The first of the two integrals is linear in $\nabla\phi$ while the second is quadratic, converging to zero as the speed in traversing the path converges to zero. Therefore, the first term corresponds to the effective heat uptake in the quasi-static limit and the second corresponds to dissipation. Consequently, we define the quasi-static and dissipated heat over the interval $[0, t_f]$ as

$$\mathcal{Q}_{\text{qs}} = -k_B \int_0^{t_f} \langle T \nabla \log(\rho), \nabla \phi \rangle_\rho dt, \tag{4.29a}$$

$$\mathcal{Q}_{\text{diss}} = -\gamma \int_0^{t_f} \int \|\nabla \phi(t, x)\|^2 \rho(t, x)\, dxdt, \tag{4.29b}$$

respectively, where both expressions are path integrals on the space of probability distributions $P_2(\mathbb{R}^n)$.

From the result by Benamou and Brenier (1.1) one can infer

$$\gamma \int_0^{t_f} \int \|\nabla \phi\|^2 \rho\, dxdt \geq \frac{\gamma}{t_f} \mathrm{W}_2^2(\rho(0, \cdot), \rho(t_f, \cdot)). \tag{4.30}$$

That is, the minimum dissipation $-\mathcal{Q}_{\text{diss}}$ is precisely the Wasserstein distance between the end-point distributions suitably scaled. The bound is achieved by transporting along the geodesic with constant velocity; this can be deduced from the Cauchy-Schwarz inequality

$$\gamma \int_0^{t_f} \int \|\nabla\phi\|^2 \rho \, dx dt \geq \frac{\gamma}{t_f}\left(\int_0^{t_f} \sqrt{\int \|\nabla\phi\|^2 \rho \, dx} \, dt\right)^2,$$

which holds as an equality when the velocity remains constant along a path. The integral in parenthesis is the length of the curve $\{\rho(t,\cdot); t \in [0, t_f]\}$ in the Wasserstein-2 metric (see Sect. 1.4), and is equal to the Wasserstein-2 distance when the curve is a geodesic.

4.2.3.2 Free Energy

Having a notion of quasi-static heat, we can attempt to introduce a notion of free energy differential for the anisotropic system,

$$đ\mathcal{F} = d\mathcal{E} - đ\mathcal{Q}_{\text{qs}},$$

where instead of "$d(T\mathcal{S}_{\text{sys}})$" (now T is a matrix so this expression has no meaning), we have written it as $đ\mathcal{Q}_{\text{qs}}$, inspired by the classical definition of entropy, $dS = dQ/T$, where the heat is understood to be quasi-static. Indeed, $đ\mathcal{Q}_{\text{qs}}$ is nothing but a weighted (by T) version of $\dot{\mathcal{S}}_{\text{sys}} = -\int \langle \nabla \log \rho, v\rho \rangle dx$ (Eq. (3.8)). Using the first law

$$đ\mathcal{F} = đ\mathcal{W} + đ\mathcal{Q} - đ\mathcal{Q}_{\text{qs}} = đ\mathcal{W} - đ\mathcal{W}_{\text{diss}},$$

where we have used the fact that $\mathcal{Q}_{\text{qs}} = \mathcal{Q} + \mathcal{W}_{\text{diss}}$. Integrating over time one can now recover the second law in terms of energy (c.f. (2.16) and (2.18))

$$\int_0^{t_f} đ\mathcal{F} = \mathcal{W} - \gamma \int_0^{t_f} \int \|\nabla\phi\|^2 \rho dx dt \leq \mathcal{W} - \frac{\gamma}{t_f} \mathrm{W}_2(\rho_0, \rho_{t_f})^2. \tag{4.31}$$

Therefore, this definition of free energy differential not only satisfies the principle of minimal work (4.31), but also vanishes at steady-state (since both $d\mathcal{E}$ and $đ\mathcal{Q}_{\text{qs}}$ do).

However, the quasi-static heat differential $đ\mathcal{Q}_{\text{qs}}$ is an imperfect differential, its integral depends on the path taken and not only on the endpoints. Therefore, $\mathcal{Q}_{\text{qs}}$ is not a function of the state of the system, and nor is $\mathcal{F}$. A possible way around this problem can be found by noting that $\mathcal{Q}_{\text{qs}}$ is a line integral; it only depends on the path taken between any two given points and not on how this path is parameterized. Hence, fixing a path between two end-points uniquely determines the quasi-static heat between them. A natural choice for this task is the 2-Wasserstein geodesic, i.e., the path of minimum dissipation. With this choice, if the end-points are equilibrium states, we recover the classical definition of free energy difference.[4]

4.2.3.3 Geometry of Cyclic Processes

Let us consider a cyclic transition of period t_f. From the first law of thermodynamics, the work output is the difference between the quasi-static heat and the dissipation, i.e.,

$$\mathcal{W}_{\text{out}} = \mathcal{Q}_{\text{qs}} - \mathcal{W}_{\text{diss}},$$

where $\mathcal{W}_{\text{diss}} = -\mathcal{Q}_{\text{diss}}$. Moreover, we define the efficiency of the cycle as the ratio between the work output and the maximum amount of work that can be extracted in a quasi-static setting [15], this is,

$$\eta = \frac{\mathcal{W}_{\text{out}}}{\mathcal{Q}_{\text{qs}}}.$$

Note that as a consequence of the second law $\eta \leq 1$. We can now cast the problem of maximizing work output over a cycle through potential control as follows:

Problem of Maximizing Work Output
Consider the n-dimensional overdamped Langevin system (4.23) with U arbitrary. Determine

$$\max\{\,\mathcal{W}_{\text{out}} \mid \partial_t \rho + \nabla \cdot (\rho \nabla \phi) = 0\,\} \tag{4.32}$$

over all closed paths $\{\rho(t,\cdot) \mid t \in [0, t_f]$, with $\rho(0,\cdot) = \rho(t_f,\cdot)\}$ on the thermodynamic manifold $P_2(\mathbb{R}^n)$.

Necessary conditions for optimality can be readily obtained by considering the Lagrangian

[4] This can be seen from the fact that 2-Wasserstein geodesics between product states retain the product structure. Hence, the different degrees of freedom can be treated independently, and the free energy can be written as a summation of the free energies of the independent degrees of freedom, each of which is subjected to a unique, constant temperature. More on this structure can be found in the proof of (5.26).

$$\mathcal{J} = \int_0^{t_f} \int \Big(k_B (T \nabla \log(\rho))' \nabla\phi \, \rho + \gamma \|\nabla\phi\|^2 \rho + \lambda(\partial_t \rho + \nabla \cdot (\rho \nabla\phi)) \Big) dx dt,$$

with $\lambda(t, x)$ a Lagrange multiplier. Its first variation is

$$\begin{aligned} \delta\mathcal{J} = \int_0^{t_f} \int_{\mathbb{R}^n} \Big(& - \nabla \cdot \left(\rho \left[k_B T \nabla \log(\rho) + 2\gamma \nabla\phi - \nabla\lambda \right]\right) \delta\phi \\ & + \Big[- k_B \nabla \cdot (T' \nabla\phi) + \gamma \|\nabla\phi\|^2 - \partial_t \lambda - (\nabla\phi)' \nabla\lambda \Big] \delta\rho \\ & + \Big[\partial_t \rho + \nabla \cdot (\rho \nabla\phi) \Big] \delta\lambda \Big) dx dt, \end{aligned}$$

where we have used integration by parts. The first-order necessary conditions for optimality are obtained by setting the variations to zero, namely

$$\begin{aligned} \nabla\phi &= \frac{1}{2\gamma} \left(\nabla\lambda - k_B T \nabla \log(\rho) - \chi \right), \\ \partial_t \lambda &= -(\nabla\phi)' \nabla\lambda + \gamma \|\nabla\phi\|^2 - k_B \nabla \cdot (T' \nabla\phi), \\ \partial_t \rho &= -\nabla \cdot (\rho \nabla\phi), \end{aligned}$$

where χ is a divergence-free vector-field (with respect to ρ), ensuring $\nabla\phi$ is of gradient form.

4.2.3.4 Isoperimetric Problem

We consider the problem (4.32) viewing trajectories as evolving on a two-dimensional submanifold $\mathcal{M} \subset P_2(\mathbb{R}^n)$ with coordinates (λ_1, λ_2); a similar two-dimensional problem has been previously studied in the linear response regime [29]. Thus, we view (λ_1, λ_2) as the two controlled degrees of freedom, while the state of our system $\rho(\lambda_1(t), \lambda_2(t), x)$, $t \in [0, t_f]$, traverses a closed path on $\mathcal{M}$ encircling a domain $\mathcal{D} \subset \mathcal{M}$ and tracing its boundary $\partial\mathcal{D}$. In order to ensure that the considered paths $\{\rho(\lambda_1(t), \lambda_2(t), x) \,:\, t \in [0, t_f]\}$ are realizable, this is, that there exists a ∇U that steers the thermodynamic system (4.24) along the desired curve $\{\rho(\lambda_1(t), \lambda_2(t), x) \,:\, t \in [0, t_f]\}$, we only consider curves that satisfy[5] assumption A1 at all times, and for which $\|\frac{\dot\rho}{\rho}\|_\rho^2 < \infty$, where $\dot\rho = \partial_t \rho$, for all $t \in [0, t_f]$.

When $\partial\mathcal{D}$ is traversed with constant velocity by the curve $\{\alpha(t) = (\lambda_1(t), \lambda_2(t)) \mid t \in [0, t_f]\}$, the quasi-static heat (4.29a) can be written as an area integral, and the dissipation (the negative of (4.29b)) expressed as a function of the length of the

[5] This ensures the existence of a unique U that solves the Poisson equation $\nabla \cdot (\rho(\nabla U + k_B T \nabla \log \rho)) = \gamma \dot\rho$.

curve, i.e.

$$\mathcal{Q}_{\text{qs}} = k_B T_r \mathcal{A}_w, \qquad \mathcal{W}_{\text{diss}} = \frac{\gamma \ell_c^2}{t_f} \ell^2, \tag{4.34}$$

where we have defined the weighted area of $\mathcal{D}$ and its perimeter by

$$\mathcal{A}_w = \frac{1}{T_r} \oint_{\partial \mathcal{D}} -\left\langle T \nabla(\log(\rho)), \frac{\nabla \phi_t}{\|\nabla \phi\|_\rho} \right\rangle_\rho ds,$$
$$\ell = \frac{1}{\ell_c} \oint_{\partial \mathcal{D}} ds,$$

where

$$\|\nabla \phi\|_\rho = \left(\int \|\nabla \phi\|^2 \rho dx \right)^{1/2},$$

and $ds = \|\nabla \phi\|_\rho dt$ represents the differential of the arc-length in $\mathcal{M} \subset P_2(\mathbb{R}^n)$ metrized with the Wasserstein W$_2$ metric; T_r and ℓ_c represent reference temperature and characteristic length, making the area and the length dimensionless quantities.

In terms of our two parameters, these two integrals can be written as

$$\mathcal{A}_w = \iint_{\mathcal{D}} w(\lambda_1, \lambda_2) \sqrt{\det(g_W)} d\lambda_1 d\lambda_2,$$
$$\ell = \frac{1}{\ell_c} \oint_{\partial \mathcal{D}} ds,$$

where the differential of the arc-length can now expressed as $ds = \|\dot{\alpha}\|_{g_W} dt$, with the 2-Wasserstein norm

$$\|\dot{\alpha}\|_{g_W} := (\dot{\alpha}' g_W \dot{\alpha})^{1/2}$$

of the velocity of the curve α expressed with a quadratic form using the matrix g_W. Thence, $w(\lambda_1, \lambda_2)$ is a work-density relative to the canonical 2-form $\sqrt{\det(g_W)} d\lambda_1 d\lambda_2$, that can be obtained using Stokes' theorem.

Consequently, we can write the work output $\mathcal{W}_{\text{out}}$ and efficiency η in terms of the area and length as

$$\mathcal{W}_{\text{out}} = k_B T_r \big(\mathcal{A}_w - \mu \ell^2 \big) \quad \text{and} \quad \eta = 1 - \mu \frac{\ell^2}{\mathcal{A}_w}, \tag{4.35}$$

where $\mu = \frac{t_c}{t_f}$ is a dimensionless constant, with $t_c = \frac{\gamma \ell_c^2}{k_B T_r}$ the characteristic time that a Brownian motion with intensity $\sqrt{\gamma^{-1} k_B T_r}$ needs to traverse a distance ℓ_c on average. Thus, the problem of maximizing work output over a cycle on the manifold of thermodynamic states can be recast as

$$\mathcal{W}^*(\mu) := k_B T_r \max_{\mathcal{D}} \{\mathcal{A}_w - \mu \ell^2\}, \tag{4.36}$$

for different values of μ. Maximization of $\mathcal{A}_w - \mu\ell^2$ relates to the isoperimetric problem, which is that of maximizing area for a fixed length of its perimeter, this is:

Isoperimetric Problem
Consider the manifold $\mathcal{M}$ equipped with metric g_W. Determine

$$\mathcal{A}_w^*(\ell) := \max\{\mathcal{A}_w \mid \ell \text{ is specified}\}, \tag{4.37}$$

over all closed paths encircling a domain $\mathcal{D} \subset \mathcal{M}$.

One can understand the equivalence between (4.36) and the isoperimetric problem in (4.37) by viewing μ in (4.36) as a Lagrange multiplier for the isoperimetric problem (4.37), where the term $\mu\ell^2$ can be viewed as a penalty on the length ℓ. Then, it is clear that the larger the length ℓ, the smaller the corresponding parameter μ, and vice-versa. Similarly to the previous section, we can obtain a first-order condition for optimal cycles through variational analysis, this is,

$$\frac{\kappa(\lambda_1, \lambda_2)}{w(\lambda_1, \lambda_2)} = \frac{1}{2\ell\mu}. \tag{4.38}$$

The solution to the isoperimetric problem, $\mathcal{A}_w^*(\ell)$, helps answer a variety of questions on optimizing control strategies. Specifically, the maximal work output

$$\mathcal{W}^*(\mu) = k_B T_r \max_{\ell}\{\mathcal{A}_w^*(\ell) - \mu\ell^2\}$$

takes place where $d\mathcal{A}_w^*(\ell)/d\ell^2 = \mu$. Also, $\mathcal{A}_w^*(\ell)$ allows computing the maximal work for a given efficiency η

$$\mathcal{W}^*(\mu)|_\eta = k_B T_r \eta \max_{\ell} \left\{ \mathcal{A}_w^*(\ell) \mid \eta = 1 - \mu \frac{\ell^2}{\mathcal{A}_w^*(\ell)} \right\},$$

which is obtained for ℓ such that $\mathcal{A}_w^*(\ell) = \frac{\mu}{1-\eta}\ell^2$. The existence of these operating points is guaranteed if $\lim_{\ell\to\infty} \mathcal{A}_w/\ell^2 = 0$, which is satisfied for manifolds with strictly negative curvature, as we will see in the following.

Even if the relationship to the isoperimetric problem is particular to two dimensions, it should be pointed out that, given (4.29a) and (4.29b), similar interpretations may be available higher/infinite-dimensional cases. Specifically, the line integral in (4.29a) may define a notion of area of a minimal surface enclosed by the arbitrary-dimensional curve (with square-length still given by (4.29b)). However, generalizing the presented results to higher dimensions is much more technical and left for future work.

4.2.3.5 Isoperimetric Inequality

Let us now take a look back at the expression for efficiency in terms of area and length (4.35). The problem of bounding efficiency is equivalent to that of finding an isoperimetric inequality in the space of thermodynamic states.

Isoperimetric inequalities, for a simply-connected domain D with area A and length l on a manifold with Gaussian curvature G, take the form [38]

$$l^2 \geq 4\pi A - 2\left(\iint_D G^+\right) A,$$

$$l^2 \geq 4\pi A - (\sup_{\mathcal{D}} G) A^2,$$

among others, where $G^+(p) = \max\{G(p), 0\}$. We illustrate how isoperimetric inequalities can be used to bound efficiency by specializing to the case when our manifold of thermodynamic states has negative curvature. Indeed, if $G \leq 0$, then $\ell^2 \geq 4\pi\mathcal{A}$, where $\mathcal{A} = \iint_{\mathcal{D}} \sqrt{\det(g_W)} d\lambda_1 \lambda_2$ is the area of $\mathcal{D}$ with respect to the canonical 2-form of $\mathcal{M}$. Thus, we obtain the following result:

Isoperimetric Inequality
Consider a two-dimensional submanifold $\mathcal{M} \subset P_2(\mathbb{R}^n)$ with coordinates (λ_1, λ_2) that has negative Gaussian curvature. Then, efficiency of a thermodynamic cycle on $\mathcal{M}$ can be bounded as follows:

$$\eta = 1 - \mu\frac{\ell^2}{\mathcal{A}_w} = 1 - \mu\frac{\ell^2}{\mathcal{A}}\frac{\mathcal{A}}{\mathcal{A}_w} \leq 1 - \frac{4\pi}{\bar{w}}\frac{t_c}{t_f},$$

where $\bar{w} = \mathcal{A}_w/\mathcal{A}$.

A conservative estimate for $\bar{w}$ is $\max_{\lambda_1,\lambda_2} w$. The above bound depends on physical parameters as well as on the chosen period, and turns negative when positive work output is not possible. In particular, the bound implies that if $t_f \leq 4\pi t_c/\bar{w}$, it is impossible to extract positive work. Thus, $4\pi t_c/\bar{w}$ constitutes a threshold for the period of work-producing cycles.

On the other hand, if the Gaussian curvature of our manifold $\mathcal{M}$ is strictly negative, that is, $G \leq -k$ for some $k > 0$, then we can conclude a tighter bound, namely, $\ell^2 \geq 4\pi\mathcal{A} + k\mathcal{A}^2$, or, in terms of $\mathcal{A}_w$,

$$\ell^2 \geq \frac{4\pi}{\bar{w}}\mathcal{A}_w + \frac{k}{\bar{w}^2}\mathcal{A}_w^2.$$

Specifically, this implies that the limit $\lim_{\ell\to\infty} \mathcal{A}_w/\ell^2 = 0$, and thus, the optimal operating points for the problem of maximizing work output and maximizing work output with a given efficiency are ensured to exist. An example of a thermodynamic manifold with negative curvature, in a slightly different setting (the linear response regime), is given in [29].

Let us conclude this chapter with an open question pertaining to the stability of the optimal periodic protocol $\{U(t,\cdot)\}_{0\leq t\leq t_f}$ that induces a periodic orbit $\{\rho(t,\cdot)\}_{0\leq t\leq t_f}$. Stability is the property of the state converging to the nominal cycle after any small perturbation. In particular, if $U(t,x) = \frac{1}{2}x'K(t)x$ is quadratic and $\rho(t,\cdot)$ is Gaussian with covariance matrix $\Sigma(t)$, we would like to ensure that a small perturbation $\Sigma(0) \to \Sigma(0) + \Delta(0)$ does not preclude the system from eventually following the nominal periodic orbit $\{\Sigma(t)\}_{0\leq t\leq t_f}$. Specifically, the perturbation from the nominal cycle obeys

$$\gamma\dot{\Delta}(t) = -K(t)\Delta(t) - \Delta(t)K(t).$$

It can be shown that $\Delta(t) \to 0$ if the integral of the smallest eigenvalue of $K(t)$ over a period is positive; this is a standard argument and relies on showing that, under the eigenvalue condition, $V(t) := \mathrm{Tr}[\Delta(t)^2]$ decreases with time (Lyapunov function). In Sect. 4.1.3 we numerically verified that the optimal curves shown in Fig. 4.4 satisfy the stated stability condition. However, providing a theoretical guarantee for the stability of all optimal curves remains open.

References

1. Cilek, V.: Earth System: History and Natural Variability, vol. 3. EOLSS Publications, Abu Dhabi (2009)
2. Battle, C., Broedersz, C.P., Fakhri, N., Geyer, V.F., Howard, J., Schmidt, C.F., MacKintosh, F.C.: Broken detailed balance at mesoscopic scales in active biological systems. Science (2016)
3. Gnesotto, F.S., Mura, F., Gladrow, J., Broedersz, C.P.: Broken detailed balance and non-equilibrium dynamics in living systems: a review. Reports on Progress in Physics. IOP publishing (2018)

4. Filliger, R., Reimann, P.: Brownian gyrator: a minimal heat engine on the nanoscale. Phys. Rev. Lett. **99**, 230602 (2007)
5. Ciliberto, S., Imparato, A., Naert, A., Tanase, M.: Statistical properties of the energy exchanged between two heat baths coupled by thermal fluctuations. J. Stat. Mech. Theory Exp. **2013**(12), P12014 (2013)
6. Dotsenko, V., Maciołek, A., Vasilyev, O., Oshanin, G.: Two-temperature Langevin dynamics in a parabolic potential. Phys. Rev. E **87**, 062130 (2013)
7. Argun, A., Soni, J., Dabelow, L., Bo, S., Pesce, G., Eichhorn, R., Volpe, G.: Experimental realization of a minimal microscopic heat engine. Phys. Rev. E **96**, 052106 (2017)
8. Fogedby, H.C., Imparato, A.: A minimal model of an autonomous thermal motor. Europhys. Lett. **119** (2017)
9. Chiang, K.-H., Lee, C.-L., Lai, P.-Y., Chen, Y.-F.: Electrical autonomous Brownian gyrator. Phys. Rev. E **96**, 032123 (2017)
10. Chang, H., Lee, C.-L., Lai, P.-Y., Chen, Y.-F.: Autonomous brownian gyrators: A study on gyrating characteristics. Phys. Rev. E **103**, 022128 (2021)
11. Movilla Miangolarra, O., Taghvaei, A., Georgiou, T.T.: A matching principle for power transfer in stochastic thermodynamics. IEEE Control Syst. Lett. (2023)
12. Movilla Miangolarra, O., Taghvaei, A., Fu, R., Chen, Y., Georgiou, T.T.: Energy harvesting from anisotropic fluctuations. Phys. Rev. E (2021)
13. Ruppeiner, G.: Riemannian geometry in thermodynamic fluctuation theory. Rev. Mod. Phys. (1995)
14. Crooks, G.E.: Measuring thermodynamic length. Phys. Rev. Lett. **99**, 100602 (2007)
15. Brandner, K., Saito, K.: Thermodynamic geometry of microscopic heat engines. Phys. Rev. Lett. **124** (2020)
16. Movilla Miangolarra, O., Taghvaei, A., Fu, R., Chen, Y., Georgiou, T.T.: Thermodynamic engine powered by anisotropic fluctuations. Phys. Rev. Res. **4**, 023218 (2022)
17. Pietzonka, P., Seifert, U.: Universal trade-off between power, efficiency, and constancy in steady-state heat engines. Phys. Rev. Lett. **120**, 190602 (2018)
18. Abdoli, I., Wittmann, R., Brader, J.M., Sommer, J.-U., Löwen, H., Sharma, A.: Tunable brownian magneto heat pump. Sci. Rep. **12**(1), 13405 (2022)
19. Lin, W., Liao, Y.-H., Lai, P.-Y., Jun, Y.: Stochastic currents and efficiency in an autonomous heat engine. Phys. Rev. E **106**, L022106 (2022)
20. Kumar, A.: Optical Feedback Traps, pp. 45–71. Springer International Publishing, Cham (2022)
21. Kumar, A., Bechhoefer, J.: Nanoscale virtual potentials using optical tweezers. Appl. Phys. Lett. **113**(18), 183702 (2018)
22. Chang, H., Chiang, K.-H., Jun, Y., Lai, P.-Y., Chen, Y.-F.: Generation of virtual potentials by controlled feedback in electric circuit systems. Phys. Rev. E **103**, 042138 (2021)
23. Ventura Siches, J., Movilla Miangolarra, O., Georgiou, T.T.: Refined bounds on energy harvesting from anisotropic fluctuations. Phys. Rev. E **109**(6), 064155 (2024)
24. Morgan, F.: Riemannian Geometry: A Beginners Guide. AK Peters, Natick (1998)
25. Abbena, E., Salamon, S., Gray, A.: Modern Differential Geometry of Curves and Surfaces with Mathematica. CRC Press, Boca Raton (2006)
26. Weinhold, F.: Metric geometry of equilibrium thermodynamics. J. Chem. Phys. (1975)
27. Zulkowski, P.R., Sivak, D.A., Crooks, G.E., DeWeese, M.R.: Geometry of thermodynamic control. Phys. Rev. E (2012)
28. Huang, Y., Krishnaprasad, P.S.: Sub-riemannian geometry and finite time thermodynamics part 1: The stochastic oscillator. Discrete Contin. Dyn. Syst.-S **13**(4), 1243 (2020)
29. Frim, A.G., DeWeese, M.R.: Geometric bound on the efficiency of irreversible thermodynamic cycles. Phys. Rev. Lett. **128**, 230601 (2022)
30. Aurell, E., Mejía-Monasterio, C., Muratore-Ginanneschi, P.: Optimal protocols and optimal transport in stochastic thermodynamics. Phys. Rev. Lett. **106** (2011)
31. Aurell, E., Gawedzki, K., Mejía-Monasterio, C., Mohayaee, R., Muratore-Ginanneschi, P.: Refined second law of thermodynamics for fast random processes. J. Stat. Phys. **147** (2012)

32. Chen, Y., Georgiou, T., Tannenbaum, A.: Stochastic control and non-equilibrium thermodynamics: fundamental limits. IEEE Trans. Autom. Control **65**(1), 252–262 (2020)
33. Dechant, A., Sakurai, Y.: Thermodynamic interpretation of Wasserstein distance. Preprint (2019). arXiv:1912.08405
34. Movilla Miangolarra, O., Taghvaei, A., Chen, Y., Georgiou, T.T.: Geometry of finite-time thermodynamic cycles with anisotropic thermal fluctuations. IEEE Control Syst. Lett. **6**, 3409–3414 (2022)
35. Huang, Y.: Optimal control of heat engines in non-equilibrium statistical mechanics. Ph.D. Thesis, University of Maryland, 2017
36. Huang, Y., Krishnaprasad, P.S.: Optimal control of a stochastic oscillator in non-equilibrium thermodynamics. In: 2016 IEEE 55th Conference on Decision and Control (CDC), pp. 197–202 (2016)
37. Alonso, P.T., Abiuso, P., Perarnau-Llobet, M., Arrachea, L.: Geometric optimization of nonequilibrium adiabatic thermal machines and implementation in a qubit system. In: PRX Quantum (2022)
38. Osserman, R.: The isoperimetric inequality. Bull. Am. Math. Soc. (1978)

Chapter 5
Minimal Entropy Production in Anisotropic Environments

Anisotropy in the environment, whether it stems from gradients of temperature, chemical concentration, or active particles, provides the fuel that feeds dynamical processes that sustain life. At the same time, anisotropy is at the core of energetic losses manifested as entropy production. In this chapter, we study these losses by considering our overdamped stochastic thermodynamic system in an anisotropic temperature heat bath. Our focus is on minimizing entropy production while driving the system between thermodynamic states in finite time [1, 2].

While entropy production in *isotropic* temperature environments can be expressed in terms of the length (in the 2-Wasserstein metric) traversed by the thermodynamic state of the system [3–6], anisotropy complicates substantially the mechanism of entropy production since, besides dissipation, seepage of energy between ambient anisotropic heat sources by way of the system dynamics is often a major contributing factor. As we will shortly see, in the presence of anisotropy, minimization of entropy production can once again be expressed via a modified Optimal Mass Transport (OMT) problem. However, in contrast to the isotropic situation that leads to a classical OMT problem and a Wasserstein length, entropy production may not be identically zero when the thermodynamic state remains unchanged (unless one has control over non-conservative forces); this is due to the fact that maintaining a NESS incurs an intrinsic entropic cost that can be traced back to a seepage of heat between heat baths.

As alluded to, NESSs represent hallmarks of life, since living matter by necessity operates far from equilibrium [7, 8]. Therefore, the question studied herein, to characterize minimal entropy production in anisotropic environments, appears of central importance in biological processes and on how such processes may have evolved to optimize for available usage of resources.

O. Movilla Miangolarra, *Stochastic Thermodynamic Treatment of Thermal Anisotropy*, Springer Theses, https://doi.org/10.1007/978-3-031-68066-3_5

5.1 Thermodynamic Space and Decomposition of Vector Fields

Thermodynamic states ρ are represented by probability distributions on $\mathbb{R}^n$, $\rho \in P_2(\mathbb{R}^n)$. In Sect. 1.4, we saw that through the definition of a suitable inner product in the tangent space of $P_2(\mathbb{R}^n)$, the space of probability distributions space admits a very rich Riemannian structure. We will now extend this theory to weighted inner products in order to account for the temperature weight in the expression of entropy production in the presence of anisotropy (c.f. (3.10)).

Our starting point is the following weighted inner product between vector fields on $\mathbb{R}^n$

$$\langle v_1, v_2\rangle_{\rho M} = \int \rho(x)\langle v_1(x), v_2(x)\rangle_M dx,$$

with induced norm $\|v\|_{\rho M} := \sqrt{\langle v, v\rangle_{\rho M}}$, where M is a symmetric positive-definite $n \times n$ matrix, and $\langle v_1(x), v_2(x)\rangle_M := v_1(x)' M v_2(x)$. The inner product defines an orthogonal decomposition

$$v = \Pi_{\rho M}\, v + \chi,$$

where $\Pi_{\rho M}$ is the projection operator given by

$$\Pi_{\rho M} v := \arg\min_{v_0}\{\|v_0\|_{\rho M} \mid v_0 = v - \chi \text{ and } \nabla \cdot (\rho\chi) = 0\}.$$

The projection is unique [9] and (under Assumptions A1 and A2 in Sect. 1.4) it is exactly of the form

$$\Pi_{\rho M} v = M^{-1}\nabla\phi, \text{ where } \phi \in H^1_\rho \text{ solves } \mathcal{L}_{\rho M^{-1}}\phi = \nabla \cdot (\rho v),$$

the (weighted) Poisson equation, with the weighted Laplacian $\mathcal{L}_{\rho M^{-1}}(\cdot) := \nabla \cdot (M^{-1}\rho\nabla(\cdot))$. This leads to the following decomposition into gradient and divergence-free parts

$$v = M^{-1}\nabla\phi + \chi, \tag{5.1}$$

analogous to the Helmholtz-Hodge decomposition [10, 11].

This decomposition can once again be used to construct the tangent space at any $\rho \in P_2(\mathbb{R}^n)$. In particular, for an admissible rate of change $\dot\rho = -\nabla \cdot (\rho v)$, induced by the vector field v, there is a corresponding gradient vector field $M^{-1}\nabla\phi = \Pi_{\rho M}\, v$, which constitutes the tangent vector. The correspondence between $\dot\rho$ and $M^{-1}\nabla\phi$ is used to equip $P_2(\mathbb{R}^n)$ with the Riemannian metric

$$\langle\dot\rho_1, \dot\rho_2\rangle_{g_W M} := \langle M^{-1}\nabla\phi_1, M^{-1}\nabla\phi_2\rangle_{\rho M}.$$

The Riemannian metric allows computing length of paths between densities. We will call the smallest distance (geodesic) between any two given densities ρ_0 and ρ_{t_f} the *weighted* Wasserstein metric $W_{2,M}(\rho_0, \rho_{t_f})$, which is given by

$$W^2_{2,M}(\rho_0, \rho_{t_f}) = t_f \min_{\rho} \int_0^{t_f} \|\dot{\rho}\|^2_{g_W M} dt, \tag{5.2}$$

subject to $\rho(0, x) = \rho_0(x)$, $\rho(t_f, x) = \rho_{t_f}(x)$.

5.2 Non-Conservative Forcing

We start by considering the problem of minimizing entropy production under the full authority of non-conservative actuation, that is, with control actuation that entails both a gradient ∇U of a potential as well as a non-zero term f in (3.1) contributing with a divergence-free component. This amounts to full control authority over the velocity vector-field v. It turns out that the minimal entropy production relates to a suitably weighted Wasserstein distance between states. This result extends the geometric characterization of entropy production in Sect. 2.3 to the case of anisotropic thermal environment.

5.2.1 *Entropy Production as a Weighted Wasserstein Length*

Consider the n-dimensional overdamped system

$$dX_t = \frac{1}{\gamma}(f(t,x) - \nabla U(t,x))dt + \sqrt{\frac{2k_B T}{\gamma}} dB_t, \tag{5.3}$$

where $X_t \in \mathbb{R}^n$ and $T = \text{diag}(T_1, T_2, \ldots, T_n)$. In Chap. 3 we showed that the total entropy production over a transition with $t \in [0, t_f]$ is given by (see (3.10))

$$\mathcal{S}_{\text{tot}} = \gamma \int_0^{t_f} \int \rho \|v\|^2_{T^{-1}} dx dt, \tag{5.4}$$

where ρ satisfies the Fokker-Planck equation

$$\partial_t \rho = -\nabla \cdot (v\rho), \tag{5.5}$$

with $v = -\frac{1}{\gamma}(\nabla U - f + k_B T \nabla \log \rho)$. By comparing this expression with the definition of the weighted Wasserstein metric (5.2) we obtain the following result:

Minimum Entropy Production Under Full Control Authority
The minimum entropy production for the overdamped system subject to anisotropic fluctuations (5.3) can be written as a weighted Wasserstein distance between states, i.e.

$$\min_{\rho,v} \mathcal{S}_{\text{tot}} = \frac{\gamma}{t_f} \mathrm{W}^2_{2,T^{-1}}(\rho_0, \rho_{t_f}), \tag{5.6}$$

where the optimization is subject to the continuity equation $\partial_t \rho + \nabla \cdot (\rho v) = 0$ together with the end-points ρ_0, ρ_{t_f} of a path $\rho(t, \cdot), t \in [0, t_f]$, effected via a control that includes gradient ∇U as well as non-conservative f component.

Note that, if a bound is imposed on the total entropy production $\mathcal{S}_{\text{tot}}$, then (5.6) provides a lower bound (speed limit) on the time needed for traversing a path that joins ρ_0 to ρ_{t_f}, namely,

$$t_f \geq \frac{\gamma \mathrm{W}^2_{2,T^{-1}}(\rho_0, \rho_{t_f})}{\mathcal{S}_{\text{tot}}}.$$

For computational purposes, it is useful to relate the weighted Wasserstein distance $\mathrm{W}_{2,M}$ to an unweighted (corresponding to the identity matrix as weight) Wasserstein distance. To this end, we first invoke the fact that an optimal transportation plan requires constancy of the velocity along paths in the Lagrangian viewpoint; this follows using the Cauchy-Schwartz inequality, as seen in Sect. 1.4. Thus, for a mass element (particle) that starts at location x and terminates at y over the time interval $[0, t_f]$, the optimal velocity remains constant and equal to

$$v(t, X_t(x)) = (y - x)/t_f$$

with the path traversed by the particular mass element being the line segment $X_t(x) = x + tv$ for $t \in [0, t_f]$. This fact turns the dynamic optimal transport (5.2) into a static (Kantorovich-type) problem, so as to be subsequently cast as an unweighted transport problem via a change of variables, as follows.

Let π be a distribution on the product space $(x, y) \in \mathbb{R}^n \times \mathbb{R}^n$ that represents the law of pairing origin x to destination y, under a transport policy. Thus, π is a coupling of random variables $X_0(x)$ and $X_{t_f}(y)$, with probability density functions $\rho_0(x)$ and $\rho_{t_f}(y)$, respectively; these are marginal distributions of π. Then,

$$\begin{aligned}\mathrm{W}^2_{2,M}(\rho_0, \rho_{t_f}) &= \min_{\pi} \int \|x - y\|^2_M d\pi(x, y) = \min_{\pi} \int \|M^{\frac{1}{2}} x - M^{\frac{1}{2}} y\|^2 d\pi(x, y) \\ &= \mathrm{W}^2_2(M^{\frac{1}{2}} \# \rho_0, M^{\frac{1}{2}} \# \rho_{t_f}) \end{aligned} \tag{5.7}$$

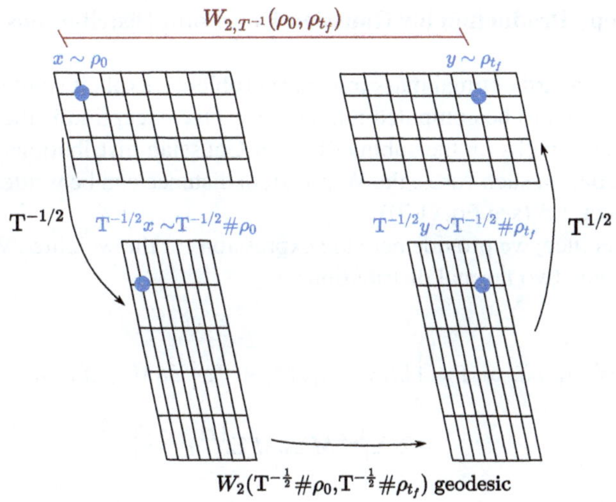

Fig. 5.1 A geometric procedure to find the optimal transportation for the weighted optimal mass transport problem

where, with a slight abuse of notation, $M^{\frac{1}{2}}\#\rho_0$ denotes the push-forward with the map[1] $g : x \mapsto M^{\frac{1}{2}}x$. Using standard theory [12], the optimal transport map for unweighted transport is given by the gradient of a convex function φ, and hence we now have $x \mapsto y = M^{-\frac{1}{2}}\nabla\varphi(M^{\frac{1}{2}}x)$; here, $\nabla\varphi$ is the optimal transport map between $M^{\frac{1}{2}}\#\rho_0$ and $M^{\frac{1}{2}}\#\rho_{t_f}$ for the unweighted cost.

Therefore, we obtain that the minimum total entropy production can be expressed as

$$\min_{\rho,v} \mathcal{S}_{\text{tot}} = \frac{1}{t_f}\frac{\gamma}{\sqrt[n]{\det(T)}}\mathrm{W}_2^2(\mathrm{T}^{-\frac{1}{2}}\#\rho_0, \mathrm{T}^{-\frac{1}{2}}\#\rho_{t_f}), \tag{5.8}$$

where $\mathrm{T} = T/\sqrt[n]{\det(T)}$ is volume preserving and the optimization is subject to the continuity equation $\partial_t\rho + \nabla\cdot(\rho v) = 0$ together with the end-point conditions $\rho(0, x) = \rho_0(x)$ and $\rho(t_f, x) = \rho_{t_f}(x)$. A geometric procedure to find the optimal transportation is thus to "warp" the space according to $\mathrm{T}^{-1/2}$, identify the optimal transport in the usual way, and then "warp" back. This procedure is illustrated in Fig. 5.1.

[1] The density of the push-forward $\rho_1 = g\sharp\rho_0$ for a differentiable $g : x \mapsto y = g(x)$ is given by $\rho_1(y) = \sum_{x\in g^{-1}(y)} \frac{\rho_0(x)}{|\det(\partial g/\partial x)|}$, this is, the measure of the pre-image, weighted by the corresponding Jacobian.

5.2.1.1 Entropy Production for Gaussian End-Point Distributions

In general, for standard optimal mass transport problems, explicit solutions are hard to come by and need to be computed numerically. One exception is the case where the transport traces paths on the submanifold of Gaussian distributions driven by a quadratic potential. In such cases, the Wasserstein distance can be written explicitly, as we saw in Sect. 1.5 (see Eq. (1.7)).

For later reference, we provide here the expression for the weighted Wasserstein-2 distance between two normal distributions:

$$\mathrm{W}_{2,M}(\rho_0, \rho_{t_f}) = \Big[\|\mu_0 - \mu_{t_f}\|_M^2 + \mathrm{Tr}\left\{\Sigma_0 M + \Sigma_{t_f} M\right. \\ \left. -2(\Sigma_{t_f}^{1/2} M \Sigma_0 M \Sigma_{t_f}^{1/2})^{1/2}\right\}\Big]^{1/2},$$

where $\rho_0 = \mathcal{N}(\mu_0, \Sigma_0)$ and $\rho_{t_f} = \mathcal{N}(\mu_{t_f}, \Sigma_{t_f})$ are Gaussian with mean μ_0 and μ_{t_f}, and covariance Σ_0 and Σ_{t_f}, respectively. The corresponding optimal trajectory (displacement interpolation) is $\{\rho(t) = \mathcal{N}(\mu(t), \Sigma(t)) \mid t \in [0, t_f]\}$, where

$$\mu(t) = \mu_0 + \frac{t}{t_f}(\mu_{t_f} - \mu_0), \tag{5.9a}$$

$$\Sigma(t) = \left(\left(1 - \frac{t}{t_f}\right)\mathrm{Id} + \frac{t}{t_f}A\right)\Sigma_0\left(\left(1 - \frac{t}{t_f}\right)\mathrm{Id} + \frac{t}{t_f}A\right)', \tag{5.9b}$$

with $A = \Sigma_{t_f}^{1/2}\left(\Sigma_{t_f}^{1/2} M \Sigma_0 M \Sigma_{t_f}^{1/2}\right)^{-1/2}\Sigma_{t_f}^{1/2} M$, and Id is the identity matrix. One can derive these results either by adjusting the argument presented in Sect. 1.5 by adding suitable weights, or from the unweighted expressions in Sect. 1.5 as explained in (5.8).

5.3 Conservative Forcing

In the vast majority of experimental set-ups, actuation, whether it comes from varying capacitance [13–15] or an optical trap [16], is conservative. That is, typically, one only has control over the gradient part of the force. In the isotropic case, this type of control actuation is sufficient for the system to follow entropy minimizing trajectories, since the control that effects optimal transport is precisely of the form $\nabla\varphi$. However, Sect. 5.2 shows that this is no longer the case in the anisotropic setting, where, in general, potential forces are no longer optimal. Thus, we now focus on minimizing entropy production when we only have access to conservative forcing. Specifically, we consider $f = 0$ and the forcing is exerted solely by varying the potential U.

As we have seen in Sect. 2.3, even if the control is so constrained, it is possible to steer the thermodynamic system between arbitrary thermodynamic states by controlling the potential function U. To see this, note that a selection of ∇U can specify the gradient part of

$$v = -\frac{1}{\gamma}(\nabla U + k_B T \nabla \log \rho), \tag{5.10}$$

and therefore, any value for $\partial_t \rho$ in (5.5). However, optimal velocities that saturate (5.6) are in general not attainable, as explained in the following.

5.3.1 Geometric Decomposition of Entropy Production

We now study the source of entropy production by identifying the contribution of the orthogonal components of the velocity field that drives the thermodynamic states. This will prove helpful in identifying optimal protocols in what follows.

Decomposition of Entropy Production

Consider a trajectory $\rho(t,\cdot) \in P_2(\mathbb{R}^n)$ that connects end-point states ρ_0, ρ_{t_f}, with $\dot{\rho} := \partial_t \rho$, and let v be a vector-field that realizes the trajectory, i.e. such that $\nabla \cdot (\rho v) + \dot{\rho} = 0$. Then, the entropy production $\gamma \int_0^{t_f} \|v\|^2_{\rho T^{-1}} dt$ can be expressed as

$$\gamma \int_0^{t_f} \|T\nabla\phi\|^2_{\rho T^{-1}} dt + \gamma \int_0^{t_f} \|\chi\|^2_{\rho T^{-1}} dt, \tag{5.11}$$

where we have used the orthogonal decomposition of v with respect to the metric $\langle \cdot, \cdot \rangle_{\rho T^{-1}}$,

$$v = T\nabla\phi + \chi, \tag{5.12}$$

where the gradient part $T\nabla\phi$ is the projection $\Pi_{\rho T^{-1}} v$, and $\nabla \cdot (\rho\chi) = 0$.

The first of these two contributions represents the minimal entropy production that is attainable when we allow non-conservative actuation to drive the system over the specified time interval between the two states (cf. Sect. 5.2)—it is precisely the Wasserstein action integral for the space equipped with the $\langle \cdot, \cdot \rangle_{g_W T^{-1}}$ Riemannian metric. Thus, it constitutes a lower bound to the total entropy production (5.11). It can be thought of as the entropic cost related to driving the thermodynamic system between the two states in finite time and will be denoted by $\mathcal{S}_{\text{ex}}$ (for *excess* cost).

The second term in (5.11) represents a contribution to the entropy production that is due to circulation in the velocity field. Such circulation, for instance, is needed to sustain a non-equilibrium steady-state (NESS). Thus, under conservative actuation, minimal entropy production can no longer be expressed in terms of a distance between end-point states, since maintaining a stationary state incurs positive entropy production. Contribution to entropy production due to circulatory currents that are generated by non-conservative forcing in a uniform heat bath [17–19] has been referred to as "housekeeping entropy production." We will follow a similar convention and label the second term in (5.11) as *housekeeping*, and denote it by $\mathcal{S}_{\rm hk}$, even if this term is not due to external driving but to the anisotropy in the environment.

Thus, the decomposition of entropy production in (5.11) can be seen as a generalization to anisotropic temperature fields of analogous decompositions discussed in earlier works [17–20]; these works consider non-conservative forcing and heat bath with uniform temperature (a single heat bath, with T scalar). In fact, the decomposition (5.11) is similar in spirit to the one recently proposed by Yoshimura et al. in [21], where they considered discrete spaces of chemical reactants that are externally driven out of equilibrium (instead of by anisotropy in the fluctuations).

5.3.1.1 Decomposition of the Circulation

The nature of $\mathcal{S}_{\rm hk}$ when the thermodynamic system is steered in the presence of thermal anisotropy is considerably more involved than when sustaining a NESS.[2] While a significant component is due to leakage of heat between the heat baths by way of the coupling between the degrees of freedom, the dynamics of the system also mediate such leakage. We make this more precise by expressing the circulation as summation of circulation due to transition and the circulation necessary to maintain a NESS. The definition of the velocity field (5.10) and its orthogonal decomposition (5.12) imply the relationship

$$v = -\frac{1}{\gamma}(\nabla U + k_B T \nabla \log \rho) = T\nabla\phi + \chi. \tag{5.13}$$

Consider a steady-state, for which $T\nabla\phi = 0$, and let $U_{\rm ss}$ and $\chi_{\rm ss}$ denote the potential function and circulation at steady-state. Then, we have the identity

$$-\frac{k_B}{\gamma} T \nabla \log \rho = \frac{1}{\gamma}\nabla U_{\rm ss} + \chi_{\rm ss},$$

[2] When maintaining a NESS ρ with two degrees of freedom (subject to "hot" and "cold" thermal excitation, respectively), $\mathcal{S}_{\rm hk} = \mathcal{Q}(\frac{1}{T_c} - \frac{1}{T_h})$, for $\mathcal{Q}$ the heat that flows from the hot to the cold heat bath, at temperatures T_h and T_c, respectively.

implying that U_{ss} and χ_{ss} are the terms in the Helmholtz-Hodge decomposition of $-\frac{k_B}{\gamma}T\nabla \log \rho$, i.e. that

$$\chi_{ss} = \frac{k_B}{\gamma}(\Pi_\rho - \mathrm{Id})T\nabla \log \rho, \tag{5.14a}$$

$$\nabla U_{ss} = -k_B \Pi_\rho T\nabla \log \rho. \tag{5.14b}$$

In general, when $T\nabla\phi \neq 0$, the potential function and circulation have additional terms due to the dynamics; these are characterized in the following statement:

Decomposition of the Circulation
Let $v = T\nabla\phi + \chi$. The circulation χ and the potential force ∇U can be decomposed into steady-state and dynamic terms as

$$\chi = \chi_{ss} + (\Pi_\rho - \mathrm{Id})T\nabla\phi, \tag{5.15a}$$

$$\nabla U = \nabla U_{ss} - \gamma\Pi_\rho T\nabla\phi, \tag{5.15b}$$

where the steady-state components are given by (5.14).

Let us now turn our attention to the characterization of states with zero circulation at steady-state, i.e. equilibrium states.

5.3.1.2 Characterization of Equilibrium States

Let us assume momentarily for simplicity that the temperatures along the different degrees of freedom are distinct, i.e., $T_i \neq T_j$ for all $i \neq j$. Entropy production at steady-state is zero if and only if $\chi_{ss} = 0$. Given (5.14a), $\chi_{ss} = 0$ implies that $T\nabla \log \rho$ is of gradient form. As a result, the orthogonality condition,

$$\langle T\nabla \log \rho, \chi_0\rangle_\rho = 0, \quad \text{for all } \chi_0 \quad \text{s.t.} \quad \nabla\cdot(\rho\chi_0) = 0,$$

must hold. Let us choose χ_0 such that $\chi_0 = \frac{1}{\rho}\Omega\nabla\psi$ for an arbitrary skew-symmetric matrix Ω and function ψ. Indeed, such a χ_0 is divergence-free since

$$\nabla\cdot(\rho\chi_0) = \nabla\cdot(\Omega\nabla\psi) = \sum_{i,j=1}^{n} \partial_i(\Omega_{ij}\partial_j\psi) = \frac{1}{2}\sum_{i,j=1}^{n}(\Omega_{ij} + \Omega_{ji})\partial_{ij}\psi = 0.$$

Integrating by parts, the orthogonality condition implies

$$\int (\nabla \log \rho)' T\Omega \nabla \psi dx = -\int \mathrm{Tr}(\Omega T \mathrm{Jac}(\log \rho))\psi dx = 0.$$

Requiring this identity to hold for any choice of ψ and Ω means that $T\mathrm{Jac}(\log \rho)$ must be symmetric, which is true iff $\partial_{ij} \log(\rho) = 0$ for all $i \neq j$. Therefore, $\log(\rho(x)) = \sum_{i=1}^{n} \log \rho_i(x_i)$, concluding that the distribution that makes the steady-state entropy production rate zero is of product form, i.e.,

$$\rho(x) = \prod_{i=1}^{n} \rho_i(x_i).$$

Conversely, if ρ is of product form, then $T\nabla \log \rho(x) = \nabla(\sum_{i=1}^{n} T_i \log \rho_i(x_i))$, and thus orthogonal to any divergence free vector-field χ_0. Therefore, we have established the following statement:

Characterization of Equilibrium Steady-States
Let $T_i \neq T_j$ for all $i \neq j$. The entropy production rate for sustaining steady-state at ρ vanishes if and only if

$$\rho(x) = \prod_{i=1}^{n} \rho_i(x_i), \tag{5.16}$$

that is, iff all degrees of freedom are independent.

If we have $T_k = T_j$ for some $k \neq j$, we can extend this result to obtain that equilibrium states $\rho(x)$ factor as $\prod_{i=1}^{m} \rho_i(\tilde{x}_i)$, where m is the number of different temperatures and $\tilde{x}_i$ is the collection of degrees of freedom with the same temperature T_i.

5.3.1.3 Directions of Vanishing Housekeeping Entropy Production

Unlike the steady-state circulation χ_{ss} which is zero only if the distribution is of product form, the total circulation χ can vanish by steering the system in specific directions. These directions result in a dynamical contribution to the circulation χ that cancels out the steady-state circulation χ_{ss} (see (5.15)). In fact, there are infinitely many such directions where the circulation vanishes:

Directions of Vanishing Housekeeping Entropy Production
A choice of potential $U(x)$ results in zero housekeeping entropy production (i.e., results in $\chi = 0$) if and only if ∇U lies in the range of $\Pi_{\rho T^{-1}}$. Specifically, if $T_i \neq T_j$ for all $i \neq j$,

$$\nabla U = -\gamma T \nabla \psi, \tag{5.17}$$

where ψ is any function of the form $\psi(x) = \sum_{i=1}^{n} \psi_i(x_i)$.

To see this, set $\chi = 0$ in (5.13) to obtain

$$\frac{1}{\gamma}\nabla U = -\frac{k_B}{\gamma} T \nabla \log \rho - T \nabla \phi.$$

Since both $(\Pi_{\rho T^{-1}} - \mathrm{Id}) T \nabla \log \rho$ and $(\Pi_{\rho T^{-1}} - \mathrm{Id}) T \nabla \phi$ are zero, $\nabla U/\gamma = -T\nabla\psi$ for a suitable ψ. It follows that if $T_i \neq T_j$ for all $i \neq j$, ψ must be such that $T\nabla\psi = \nabla\psi_T$ where $\psi_T(x) = \sum_{i=1}^{n} T_i \psi_i(x_i)$.

A potential U as in (5.17) leads to tangent directions of the form

$$T\nabla\phi = -\frac{k_B}{\gamma} T \nabla \log \rho + T \nabla \psi, \tag{5.18}$$

having vanishing housekeeping entropy production. Among such directions, it is interesting to characterize the one with minimum excess entropy production:

$$\min \dot{\mathcal{S}}_{\mathrm{ex}} = \min_{\nabla\phi} \gamma \|T\nabla\phi\|^2_{\rho T^{-1}}, \quad \text{s.t.} \quad \chi = 0. \tag{5.19}$$

Indeed, the formula for tangent directions with vanishing circulation (5.18) implies

$$\gamma \|T\nabla\phi\|^2_{\rho T^{-1}} = \gamma \left\|\frac{k_B}{\gamma} T \nabla \log \rho - T\nabla\psi\right\|^2_{\rho T^{-1}} = \frac{k_B^2}{\gamma} \|\nabla \log \rho - \nabla\widetilde{\psi}\|^2_{\rho T},$$

with $\widetilde{\psi} = k_B^{-1}\gamma\psi$. Minimizing the squared-norm

$$\|\nabla \log \rho - \nabla\widetilde{\psi}\|^2_{\rho T} = \sum_{i=1}^{n} T_i \int \left(\partial_i \log \rho(x) - \partial_i \widetilde{\psi}_i(x_i)\right)^2 \rho(x) dx,$$

involves n independent minimizations, each over $\partial_i \widetilde{\psi}_i(x_i)$ for $i = 1, 2, \ldots, n$. The solution to each minimization problem is the conditional expectation[3]

$$\partial_i \widetilde{\psi}_i(x_i) = \int \partial_i \log \rho(x) \rho(x_{/i}|x_i)\, dx_{/i},$$

where $x_{/i}$ denotes all coordinates except x_i and $\rho(x_{/i}|x_i)$ is the conditional density of $x_{/i}$ given x_i. Upon using the definition of conditional density, $\rho(x_{/i}|x_i) = \rho(x)/\rho_i(x_i)$ where $\rho_i(x_i) = \int \rho(x)dx_{/i}$ is the i-th marginal of ρ, we obtain

$$\partial_i \widetilde{\psi}_i(x_i) = \int \frac{\partial_i \rho(x)}{\rho(x)} \frac{\rho(x)}{\rho_i(x_i)} dx_{/i} = \frac{1}{\rho_i(x_i)} \partial_i \int \rho(x) dx_{/i} = \partial_i \log \rho_i(x_i).$$

Therefore, if $T_i \neq T_j$ for all $i \neq j$, the minimum excess entropy production *rate* with zero circulation,

$$\frac{k_B^2}{\gamma} \|\nabla \log \rho - \nabla \log \bar{\rho}\|_{\rho T}^2, \tag{5.20}$$

is attained using $\nabla U = -\gamma T \nabla \psi$ with

$$\psi = \frac{k_B}{\gamma} \sum_{i=1}^{n} \log \rho_i(x_i). \tag{5.21}$$

Here, $\bar{\rho}(x)$ is the density of the independent distribution with the same marginals as ρ, i.e. $\bar{\rho}(x) = \prod_{i=1}^{n} \rho_i(x_i)$. Vanishing minimum excess entropy production in (5.20) is achieved if and only if ρ is of product form, i.e., as in Eq. (5.16). This stands to reason, since $\chi_{ss} \neq 0$ implies that the only way to make χ vanish in (5.15) is through a non-zero velocity $T\nabla\phi$, leading to non-zero excess entropy.

5.3.2 *Direction of Least Entropy Production*

Leaving trajectories with vanishing housekeeping entropy production aside, directions minimizing total entropy production rate can be identified. These are of interest, not only for standard energetic considerations, but also because gradient flows that minimize entropy rate are envisioned to have physical and biological significance. Thus, below we identify the potential U that minimizes instantaneous entropy production *rate*.

---

[3] By definition of the conditional expectation as the least-mean-square estimator [22, Prop. 2.3.3].

Direction of Least Entropy Production
The vector-field ∇U that minimizes the entropy production *rate* $\dot{\mathcal{S}}_{\text{tot}}$ is given in terms of the projection of $-k_B \nabla \log(\rho)$ according to

$$\nabla U = -k_B T \Pi_{\rho T}(\nabla \log(\rho)). \tag{5.22}$$

Hence, the potential function $U \in H^1(\rho)$ is the unique solution to the Poisson equation

$$\mathcal{L}_{\rho T^{-1}} U = -k_B \Delta \rho, \quad \text{where } \mathcal{L}_{\rho M}(\cdot) := \nabla \cdot (\rho M \nabla(\cdot)). \tag{5.23}$$

Indeed, using the decomposition

$$k_B \nabla \log \rho = T^{-1} \nabla \psi + \chi,$$

where $T^{-1}\nabla\psi = k_B \Pi_{\rho T} \nabla \log \rho$ and $\nabla \cdot (\rho\chi) = 0$, the entropy production rate

$$\begin{aligned}\gamma \dot{\mathcal{S}}_{\text{tot}} &= \|\nabla U + k_B T \nabla \log \rho\|^2_{\rho T^{-1}} = \|T^{-1}\nabla U + k_B \nabla \log \rho\|^2_{\rho T} \\ &= \|T^{-1}\nabla U + T^{-1}\nabla\psi\|^2_{\rho T} + \|\chi\|^2_{\rho T}\end{aligned}$$

is minimized by $\nabla U = -\nabla\psi = -k_B T \Pi_{\rho T} \nabla \log \rho$. The Poisson equation follows from the definition of the projection and the fact that $\nabla \cdot (\rho\chi) = 0$. Existence and uniqueness of a (weak) solution to the Poisson equation follow from the Poincaré inequality and finiteness of $\|\frac{\Delta\rho}{\rho}\|^2_{\rho}$, which hold under Assumption A1 [23].

5.3.3 *Minimal Entropy Production Between End-Points*

We now consider minimizing entropy production along *a path* between two distributions driven by conservative actuation. As we have mentioned, the key technical difficulty lies in the control being constrained. This prevents us from eliminating the housekeeping part of the entropy production and minimizing the excess part (as in Sect. 5.2). Due to the systemic nature of the housekeeping part, the problem of minimizing entropy production becomes a challenging trade-off between housekeeping and excess entropies, which are coupled by the ∇U term that enters in both.

To tackle this problem, let us first rewrite the rate of entropy production as

$$\begin{aligned}\dot{\mathcal{S}}_{\text{tot}} =& \frac{1}{\gamma}\int \|\nabla U + k_B T\nabla \log \rho\|^2_{T^{-1}}\rho dx\\ =& \frac{1}{\gamma}\Big[\int \|\nabla U\|^2_{T^{-1}}\rho dx - k_B^2\int \|\nabla \log\rho\|^2_T \rho dx\Big] + 2\dot{\mathcal{S}}_{\text{sys}},\end{aligned}$$

using that $\dot{\mathcal{S}}_{\text{sys}} = -k_B\int \partial_t\rho \log\rho dx = -k_B\int(\nabla\log\rho)' v\rho dx$ via integration by parts. Since $\int_0^{t_f}\dot{\mathcal{S}}_{\text{sys}}dt = \mathcal{S}_{\text{sys}}(\rho(t_f,\cdot)) - \mathcal{S}_{\text{sys}}(\rho(0,\cdot))$ only depends on the end-point distributions, minimizing entropy production over the transition amounts to solving

$$\min_{U,\rho}\frac{1}{\gamma}\int_0^{t_f}\int\Big[\|\nabla U\|^2_{T^{-1}} - k_B^2\|\nabla\log\rho\|^2_T\Big]\rho dx dt, \tag{5.24}$$

subject to the continuity Eq. (5.5) and the end-point conditions.

To solve this optimization problem it is useful to consider the augmented Lagrangian,

$$\begin{aligned}\mathcal{J} =& \frac{1}{\gamma}\int_0^{t_f}\int\Big[(\nabla U)'T^{-1}\nabla U - k_B^2(\nabla\log\rho)'T\nabla\log\rho\Big]\rho dx dt\\ &+\int_0^{t_f}\int\lambda\Big[\partial_t\rho - \frac{1}{\gamma}\nabla\cdot((\nabla U + k_B T\nabla\log\rho)\rho)\Big]dx dt,\end{aligned}$$

with λ a Lagrange multiplier. Its first variation can be written as

$$\begin{aligned}\delta\mathcal{J} = \frac{1}{\gamma}\int_0^{t_f}\int\Big[&2(\nabla\delta_U)'T^{-1}\nabla U\rho + (\nabla U)'T^{-1}\nabla U\delta_\rho - 2k_B^2(\nabla\tfrac{\delta_\rho}{\rho})'T\nabla\log\rho\rho\\ &- k_B^2(\nabla\log\rho)'T\nabla\log\rho\delta_\rho + \delta_\lambda\big(\gamma\partial_t\rho - \nabla\cdot((\nabla U + k_B T\nabla\log\rho)\rho)\big)\\ &+\lambda\big\{\gamma\partial_t\delta_\rho - \nabla\cdot((\nabla U + k_B T\nabla\log\rho)\delta_\rho)\\ &-\nabla\cdot((\nabla\delta_U + k_B T\nabla\tfrac{\delta_\rho}{\rho})\rho)\big\}\Big]dx dt.\end{aligned}$$

Integrating by parts and setting this to zero for all perturbations $\delta_U, \delta_\rho, \delta_\lambda$ we obtain the following first-order necessary conditions for optimality

$$\begin{aligned}\gamma\partial_t\rho &= \nabla\cdot(\rho(\nabla U + k_B T\nabla\log\rho)),\\ \gamma\partial_t\lambda &= \|\nabla U\|^2_{T^{-1}} + \tfrac{2k_B^2}{\rho}\nabla\cdot(T\nabla\rho) - k_B^2\|\nabla\log\rho\|^2_T\\ &\quad + \langle\nabla\lambda, \nabla U + k_B T\nabla\log\rho\rangle - \tfrac{1}{\rho}\nabla\cdot(\rho k_B T\nabla\lambda),\\ 0 &= \mathcal{L}_{\rho T^{-1}}U + \frac{1}{2}\nabla\cdot(\rho\nabla\lambda),\end{aligned} \tag{5.25}$$

with boundary values $\rho(0, x) = \rho_0(x)$ and $\rho(t_f, x) = \rho_{t_f}(x)$. These equations have the structure of a coupled system of partial differential equations and, in general, need to be solved numerically. However, closed-form solutions to this problem can be obtained for the following special case.

5.3.3.1 Minimal Entropy Production Between Product States

We herein focus on minimizing entropy production while transitioning between two states with mutually independent degrees of freedom, i.e.

$$\min_{U,\rho} \left\{ \mathcal{S}_{\text{tot}} \mid (5.5),\ \rho(0, x) = \prod_{i=1}^{n} \rho_i^0(x_i),\ \rho(t_f, x) = \prod_{i=1}^{n} \rho_i^f(x_i) \right\}.$$

Physically, these end-point states may represent equilibrium distributions, where entropy production vanishes. More generally, these states result from a system with uncoupled degrees of freedom—a situation in which anisotropy is virtually absent. For example, in the electrical network embodiment of Fig. 3.1a, this would entail that at times 0 and t_f the network decomposes into two circuits that do not interact, each with its own source of thermal excitation (Nyquist Johnson noise) from its own resistor at a fixed temperature.

It turns out that this problem is equivalent to minimizing excess entropy production, and that the thermodynamic state retains independence between degrees of freedom along the optimal trajectory. Thus, we circumvent the problem arising from the coupling between the excess and housekeeping entropy terms, since for this particular case excess-entropy-minimizing trajectories lead to zero housekeeping entropy.

Indeed, from the geometric decomposition in Sect. 5.3.1 we know that the minimal excess entropy production $\mathcal{S}_{\text{ex}}$, when transitioning between $\rho_0 = \prod_{i=1}^{n} \rho_i^0(x_i)$ and $\rho_{t_f} = \prod_{i=1}^{n} \rho_i^f(x_i)$, is given by

$$\min_{\rho, v} \mathcal{S}_{\text{ex}} = \frac{1}{t_f} \mathrm{W}_{2,M}^2(\rho_0, \rho_{t_f}),$$

for $M = \gamma T^{-1}$. We have that

$$\mathrm{W}_{2,M}^2(\rho_0, \rho_{t_f}) = \min_{\pi} \int \|x - y\|_M^2 d\pi(x, y),$$

where $\pi(x_1, x_2, \ldots, y_1, y_2, \ldots)$ is a coupling having marginals $\rho_i^0(x_i)$ and $\rho_i^f(y_i)$, with x_i, y_i denoting coordinates along the same i-th degree of freedom at the start and end of the transition, respectively. We note that the cost $\|x - y\|_M^2$ splits as $\sum_i^n \|x_i - y_i\|_{M_{ii}}^2$ leading to n uncoupled weighted OMT problems,

$$\mathrm{W}_{2,M}^2(\rho_0,\rho_{t_f}) = \sum_i^n \min_{\pi_i} \int \|x_i - y_i\|_{M_{ii}}^2 d\pi_i(x_i, y_i),$$

where now π_i is a coupling between the starting and ending marginals of the i-th degree of freedom ρ_i^0 and ρ_i^f. Thus, the optimal velocity field along the i-th degree of freedom is of the form $T_i \partial_i \phi_i(x_i)$, concluding in an optimal velocity field for the original problem of the form $T\nabla\phi = \nabla\phi_T$ where $\phi_T(x) = \sum_{i=1}^n T_i\phi_i(x_i)$. Moreover, the resulting path of thermodynamic states $\rho(t,x)$ retains the product structure of the starting and ending marginals. This path can be realized by a control of the form $\nabla U = -\gamma T\nabla\psi$ as in Eq. (5.17), resulting in vanishing housekeeping entropy production. Therefore, the optimal protocol for minimizing excess entropy also minimizes housekeeping entropy, leading to the following result:

Minimal Entropy Production Between Product States
The solution to the problem of minimizing entropy production between product states,

$$\min_{U,\rho} \left\{ \mathcal{S}_{\text{tot}} \mid (5.5),\ \rho(0,x) = \prod_{i=1}^n \rho_i^0(x_i),\ \rho(t_f,x) = \prod_{i=1}^n \rho_i^f(x_i) \right\}, \tag{5.26}$$

coincides with the solution to the unconstrained problem (5.6) with end-points ρ_0 and ρ_{t_f} of product form as above.

5.3.4 *Control via a Quadratic Potential*

We now specialize to the case where the controlling potential is quadratic, namely,

$$U(t,x) = \frac{1}{2}x'K(t)x$$

with $x \in \mathbb{R}^n$ and $K(t) = K(t)'$. This choice of potential not only simplifies calculations, but is also extremely useful and practical, as we have seen before (e.g. Sect. 3.4.2).

Starting from an initial zero-mean Gaussian distribution, the thermodynamic state traces a path on the submanifold of Gaussian distributions

$$\rho(t,x) = \frac{1}{(2\pi)^{n/2}\det(\Sigma(t))^{1/2}} e^{-\frac{1}{2}\|x\|^2_{\Sigma(t)^{-1}}}, \tag{5.27}$$

where the covariance Σ satisfies the differential Lyapunov equation (corresponding to the Fokker-Planck equation (5.5))

$$\gamma \dot{\Sigma}(t) = -K(t)\Sigma(t) - \Sigma(t)K(t) + 2k_B T. \tag{5.28}$$

In this case, the velocity field (5.10) takes form

$$v(t,x) = -\frac{1}{\gamma}K(t)x + \frac{k_B}{\gamma}T\Sigma(t)^{-1}x, \tag{5.29}$$

linear in x, since $\nabla \log \rho(t,x) = -\Sigma(t)^{-1}x$. In the following, we specialize the results obtained in the previous sections to such linear systems.

5.3.4.1 Geometric Decomposition of Entropy Production

The geometric decomposition of entropy production of Sect. 5.3.1, specialized to the case of a Gaussian path of distributions, is as follows. Always assuming zero-mean, the path corresponds to a curve of covariance matrices $\{\Sigma(t) : t \in [0, t_f]\}$, while the entropy production is given by

$$\mathcal{S}_{\text{tot}} = \gamma \int_0^{t_f} \text{Tr}[V' T^{-1} V \Sigma] dt.$$

Here, following (5.12), $v = Vx = -\frac{1}{\gamma}\left(K - k_B T \Sigma^{-1}\right)x$ admits the orthogonal decomposition

$$Vx = \underbrace{TAx}_{T\nabla\phi} + \underbrace{\Omega\Sigma^{-1}x}_{\chi}, \tag{5.30}$$

where A is a symmetric matrix and Ω is skew-symmetric, possibly time-varying. Accordingly, thanks to the orthogonality condition $\text{Tr}[A\Omega] = 0$, the entropy production decomposes into two parts

$$\mathcal{S}_{\text{ex}} + \mathcal{S}_{\text{hk}} = \gamma \int_0^{t_f} \text{Tr}[ATA\Sigma] dt + \gamma \int_0^{t_f} \text{Tr}[\Omega' T^{-1} \Omega \Sigma^{-1}] dt, \tag{5.31}$$

in agreement with (5.11).

Echoing the development in Sect. 5.3.1, for the special case of Gaussian densities and quadratic potential, we decompose the circulation and potential into their steady-state and dynamical components. At steady-state (where $A = 0$), the decomposition (5.30) implies that

$$\frac{k_B}{\gamma}T\Sigma^{-1} - \frac{1}{\gamma}K_{ss} = \Omega_{ss}\Sigma^{-1}.$$

Using the fact that the right hand side of $K_{ss} = k_B T\Sigma^{-1} - \gamma\Omega_{ss}\Sigma^{-1}$ must be symmetric (since K_{ss} is), we obtain

$$\Omega_{ss}\Sigma^{-1} + \Sigma^{-1}\Omega_{ss} = \frac{k_B}{\gamma}(T\Sigma^{-1} - \Sigma^{-1}T),$$

and hence,

$$\Omega_{ss} = \frac{k_B}{\gamma} L_{\Sigma^{-1}}(T\Sigma^{-1} - \Sigma^{-1}T),$$

where

$$X \mapsto L_A(X) := \int_0^\infty e^{-\tau A} X e^{-\tau A'} d\tau,$$

with A a matrix with positive eigenvalues.

Similarly, since $\gamma\Omega_{ss} = k_B T - K_{ss}\Sigma$ must be skew-symmetric, we obtain

$$K_{ss} = 2k_B L_\Sigma(T).$$

When not at steady-state, $A \neq 0$ in (5.30) introduces an extra term leading to

$$\Omega = \Omega_{ss} - L_{\Sigma^{-1}}(TA - AT),$$
$$K = K_{ss} - \gamma L_\Sigma(TA\Sigma + \Sigma AT),$$

similarly to (5.15).

5.3.4.2 Directions of Vanishing Housekeeping Entropy Production

We now characterize directions with vanishing housekeeping entropy production (c.f. (5.17)), as it pertains to the Gaussian case.

In light of (5.30), for zero housekeeping entropy production ($\chi = 0$), the velocity field must be of the form $v = TAx$, with A symmetric. At the same time, $v = -\gamma^{-1}(K - k_B T\Sigma^{-1})x$. Thus K, which is symmetric, must be such that $T^{-1}K$ is also symmetric. (This argumentation recapitulates the reasoning that leads to ∇U being in the range of $\Pi_{\rho T^{-1}}$.) It follows that K must be diagonal. In conclusion, the system is steered in a direction with vanishing housekeeping entropy production if and only if K is diagonal.

We can further identify choices of K that, besides ensuring $\dot{\mathcal{S}}_{\rm hk} = 0$, minimize excess entropy production *rate* as in Eq. (5.20). In the present case where $\nabla U = Kx$ with K diagonal,

$$\begin{aligned}\dot{\mathcal{S}}_{\text{ex}} &= \frac{1}{\gamma}\|k_B T\nabla \log \rho + \nabla U\|^2_{\rho T^{-1}} = \frac{1}{\gamma}\|k_B T\Sigma^{-1}x - Kx\|^2_{\rho T^{-1}} \\ &= \frac{1}{\gamma}\operatorname{Tr}(k_B^2\Sigma^{-1}T - 2k_B K + KT^{-1}K\Sigma).\end{aligned}$$

This is minimized for $K_{ii} = k_B T_i(\Sigma_{ii})^{-1}$, in agreement with (5.21).

5.3.4.3 Direction of Least Entropy Production Rate

We can readily obtain the potential function (i.e., the gain $K(t)$) that minimizes the entropy production *rate*. Indeed, with $U(x) = x'Kx/2$ and Gaussian distribution (5.27), Eq. (5.23) translates into

$$T^{-1}K\Sigma + \Sigma K T^{-1} = 2k_B I,$$

or, equivalently,

$$K\Sigma T + T\Sigma K = 2k_B T^2,$$

with unique solution

$$K = 2k_B L_{T\Sigma}(T^2),$$

where we highlight that $T\Sigma$ is a matrix with positive eigenvalues, so the integral $L_{T\Sigma}(T^2)$ converges.

5.3.4.4 Minimal Entropy Production Between End-Points

Next, we specialize the first-order optimality condition (5.25) for *transition between end-point states* to Gaussian states and transition path. We adopt the ansatz that the Lagrange multiplier is of the form

$$\lambda(t, x) = \frac{1}{2}x'\Lambda(t)x + c(t).$$

Then, the optimal K, Σ and Λ must satisfy

$$\begin{aligned}\gamma\dot{\Sigma} &= -K\Sigma - \Sigma K + 2k_B T, \\ \gamma\dot{\Lambda} &= \Lambda K + K\Lambda + 2KT^{-1}K + 2k_B^2\Sigma^{-1}T\Sigma^{-1}, \\ \gamma\dot{c} &= -2k_B^2 \operatorname{Tr}(T\Sigma^{-1}) - k_B \operatorname{Tr}[T\Lambda],\end{aligned}$$

$$K = -\frac{1}{2}L_{T\Sigma}(T\Lambda\Sigma T + T\Sigma\Lambda T),$$

translating (5.25) to the quadratic actuation case. This is a set of coupled algebraic-differential equations with two-point boundary conditions, that have to be solved numerically (e.g., via a shooting method).

5.3.4.5 Minimal Entropy Production Between Gaussian Product States

We now focus on minimizing entropy production while transitioning between two Gaussian states with mutually independent degrees of freedom, i.e., with diagonal covariances. As in Sect. 5.3.3.1, this is equivalent to solving the unconstrained problem (5.6), subject to a quadratic potential with end-points $\rho_0 = \mathcal{N}(0, \Sigma_0)$ and $\rho_{t_f} = \mathcal{N}(0, \Sigma_{t_f})$ with Σ_0 and Σ_{t_f} diagonal. To see this, note that the optimal solution to (5.6) with Gaussian end-points is given by the Gaussian interpolation (5.9), where $\Sigma(t)$ remains diagonal at all times, as Σ_0 and Σ_{t_f} are diagonal (recall that M in (5.9b) is to be replaced by T^{-1}, which is diagonal). Consequently, $K(t)$ in (5.28) must also be diagonal. Since $\Sigma(t)$ and $K(t)$ are diagonal, the matrix $\gamma^{-1}(-K(t) + k_B T\Sigma(t)^{-1})$ that defines the velocity field is also diagonal. Hence, the velocity field has no circulation ($\Omega = 0$) and thus, the housekeeping entropy production vanishes. Evidently, this is due to the fact that T, K, and Σ are "aligned," or in other words, they commute.

Unfortunately, the above considerations tell us very little about optimal trajectories that start or end at non-equilibrium steady-states. Next, we provide insight into such a scenario for a two-dimensional setting.

5.3.5 The Brownian Gyrator Case

We now consider that $n = 2$, where we may picture a particle with two degrees of freedom subject to respective stochastic excitations that correspond to temperatures $T_1 > T_2$. This is precisely the setting of the Brownian Gyrator already discussed in Sects. 3.4 and 4.1 in the context of energy extraction. The goal of the present section is to characterize entropy production, resulting from heat transfer as well as the dynamics in a non-stationary setting, and to determine time-varying control that precisely minimizes this entropy production.

5.3.5.1 Explicit Expressions of $\mathcal{S}_{\mathrm{ex}}$ and $\mathcal{S}_{\mathrm{hk}}$

Starting from (5.31), the total entropy production can be expressed as

$$\underbrace{\gamma\int_0^{t_f}\mathrm{Tr}[ATA\Sigma]dt}_{\mathcal{S}_{\mathrm{ex}}}+\underbrace{\gamma\int_0^{t_f}\mathrm{Tr}[\Sigma^{1/2}T^{-1}\Sigma^{1/2}\widetilde{\Omega}\widetilde{\Omega}']dt}_{\mathcal{S}_{\mathrm{hk}}},$$

where $\widetilde{\Omega}=\Sigma^{-1/2}\Omega\Sigma^{-1/2}$ is still a skew-symmetric matrix. Since we have defined our decomposition (5.30) so that

$$V\Sigma=-\gamma^{-1}K\Sigma+\gamma^{-1}k_BT=TA\Sigma+\Omega,$$

we write

$$\gamma V\Sigma+\gamma\Sigma V'=-K\Sigma-\Sigma K+2k_BT=\gamma TA\Sigma+\gamma\Sigma AT.$$

Using this expression, the Lyapunov equation (5.28) becomes

$$\begin{aligned}\dot{\Sigma}&=TA\Sigma+\Sigma AT,\ \text{or,}\\ \dot{\widetilde{\Sigma}}&=\widetilde{A}\widetilde{\Sigma}+\widetilde{\Sigma}\widetilde{A},\end{aligned}\tag{5.34}$$

for $\widetilde{A}:=T^{1/2}AT^{1/2}$ and $\widetilde{\Sigma}:=T^{-1/2}\Sigma T^{-1/2}$.

The excess entropy production can now be written as

$$\mathcal{S}_{\mathrm{ex}}=\frac{\gamma}{2}\int_0^{t_f}\mathrm{Tr}[\dot{\Sigma}A]dt=\frac{\gamma}{2}\int_0^{t_f}\mathrm{Tr}[\dot{\widetilde{\Sigma}}\widetilde{A}]dt=\frac{\gamma}{2}\int_0^{t_f}\mathrm{Tr}[L_{\widetilde{\Sigma}}(\dot{\widetilde{\Sigma}})\dot{\widetilde{\Sigma}}]dt.$$

To obtain the last equality above we have used the identity $\widetilde{A}=L_{\widetilde{\Sigma}}(\dot{\widetilde{\Sigma}})$ as the solution to (5.34). Note that this expression for $\mathcal{S}_{\mathrm{ex}}$ is identical to the expression for the dissipation in Sect. 4.1 (c.f. Eq. (4.9b)), but for the fact that instead of Σ we now have $\widetilde{\Sigma}$. This is not a coincidence, since by changing $\Sigma\to\widetilde{\Sigma}$ we are simply accounting for the extra T^{-1} weight of the inner product that describes entropy production (see (5.4)), in analogy with Eq. (5.7), where we changed variables from x to $T^{-1/2}x$ in order to write entropy production in terms of an unweighted Wasserstein-2 distance.

To simplify calculations we introduce the same parameterization as in Sect. 4.1 (see Eq. (4.10)), now for $\widetilde{\Sigma}$ instead of Σ, this is,

$$\widetilde{\Sigma}(r,\theta)=R\Big(-\frac{\theta}{2}\Big)\sigma^2(r)R\Big(\frac{\theta}{2}\Big),\tag{5.35}$$

where

$$R(\vartheta)=\begin{bmatrix}\cos(\vartheta)&\sin(\vartheta)\\-\sin(\vartheta)&\cos(\vartheta)\end{bmatrix}\ \text{and}\ \sigma^2(r)=\frac{\ell_c^2}{\sqrt{T_1T_2}}\begin{bmatrix}e^r&0\\0&e^{-r}\end{bmatrix},$$

are matrices, orthogonal and diagonal, respectively, and where $\ell_c = \sqrt[4]{\det(\Sigma(t))}$ is the (constant) characteristic length of the system.

With this parameterization, one can explicitly express the excess part of entropy production as (see Sect. 4.1.3, Eq. (4.11) for the detailed computation)

$$\mathcal{S}_{\text{ex}} = k_B \tau \int_0^{t_f} \left(\cosh(r)\dot{r}^2 + \sinh(r)\tanh(r)\dot{\theta}^2 \right) dt, \tag{5.36}$$

where $\tau = \gamma \ell_c^2 / (2k_B\sqrt{T_1 T_2})$ is a *characteristic time* constant in that it is the average time that a Brownian motion with intensity $\sqrt{2\gamma^{-1}k_B\sqrt{T_1T_2}}$ needs to traverse a distance ℓ_c. Note that (5.36) is quadratic in the velocities $(\dot{r}, \dot{\theta})$ and therefore vanishes as $t_f \to \infty$. It is precisely the *weighted* Wasserstein action integral, expressed in terms of r and θ.

Let us look back at the housekeeping term and define ω through

$$\widetilde{\Omega} = \omega\hat{\Omega} \text{ with } \hat{\Omega} = \begin{bmatrix} 0 & 1 \\ -1 & 0 \end{bmatrix}.$$

Thereby, $\widetilde{\Omega}\widetilde{\Omega}' = \omega^2 I$ and the housekeeping entropy production can be written as

$$\begin{aligned} \mathcal{S}_{\text{hk}} &= \gamma \int_0^{t_f} \omega^2 \operatorname{Tr}[\Sigma^{1/2}T^{-1}\Sigma^{1/2}]dt \\ &= \gamma \int_0^{t_f} \omega^2 \operatorname{Tr}[\widetilde{\Sigma}]dt = \frac{2\gamma\ell_c^2}{\sqrt{T_1T_2}} \int_0^{t_f} \omega^2 \cosh(r)dt, \end{aligned}$$

where ω is to be determined so that

$$K = -\gamma T A - \omega\gamma\Sigma^{1/2}\hat{\Omega}\Sigma^{-1/2} + k_B T\Sigma^{-1}$$

is symmetric. Imposing $\operatorname{Tr}[K\hat{\Omega}] = 0$, which is equivalent to K being symmetric, we obtain

$$\begin{aligned} &\operatorname{Tr}[T(-\gamma A + k_B\Sigma^{-1})\hat{\Omega}] - \gamma\omega \operatorname{Tr}[\Sigma^{1/2}\hat{\Omega}\Sigma^{-1/2}\hat{\Omega}] = 0 \\ \implies &\omega = \frac{\operatorname{Tr}[T(k_B\Sigma^{-1} - \gamma A)\hat{\Omega}]}{\gamma \operatorname{Tr}[\Sigma^{1/2}\hat{\Omega}\Sigma^{-1/2}\hat{\Omega}]}. \end{aligned}$$

The numerator can be written in terms of $\widetilde{\Sigma}$ and $\dot{\widetilde{\Sigma}}$ as

$$\operatorname{Tr}[T(k_B\Sigma^{-1} - \gamma A)\hat{\Omega}] = \operatorname{Tr}[(k_B\widetilde{\Sigma}^{-1} - \gamma L_{\widetilde{\Sigma}}[\dot{\widetilde{\Sigma}}])T^{-1/2}\hat{\Omega}T^{1/2}],$$

where we have used the fact that $\widetilde{A} = T^{1/2}AT^{1/2} = L_{\widetilde{\Sigma}}[\dot{\widetilde{\Sigma}}]$ (from Eq. (5.34)). On the other hand, diagonalizing Σ as $\Sigma = R'\Lambda R$, the denominator turns

$$\gamma\,\mathrm{Tr}[\Sigma^{1/2}\hat{\Omega}\Sigma^{-1/2}\hat{\Omega}] = \gamma\,\mathrm{Tr}[\Lambda^{1/2}\hat{\Omega}\Lambda^{-1/2}\hat{\Omega}] = -\gamma\frac{\mathrm{Tr}[\Sigma]}{\det(\Sigma)^{1/2}} = -\gamma\frac{\mathrm{Tr}[T\widetilde{\Sigma}]}{\det(T\widetilde{\Sigma})^{1/2}},$$

where the second equality follows from developing the trace on the left-hand side, and we have denoted by $\det(\Sigma)$ the determinant of Σ. Writing both numerator and denominator in terms of r and θ we obtain

$$\omega = \frac{\Delta T}{2}\frac{\left(\dot{r} + \tau^{-1}\sinh(r)\right)\sin\theta + \dot{\theta}\tanh(r)\cos\theta}{\bar{T}\cosh(r) + \Delta T\sinh(r)\cos\theta},$$

where $\Delta T := T_1 - T_2$ and $\bar{T} = T_1 + T_2$. Hence,

$$\mathcal{S}_{\mathrm{hk}} = k_B\tau\Delta T^2\int_0^{t_f}\frac{1}{\cosh(r)}\left(\frac{\sigma(r,\dot{r},\theta,\dot{\theta})}{\bar{T} + \Delta T\tanh(r)\cos\theta}\right)^2 dt, \tag{5.37}$$

with

$$\sigma(r,\dot{r},\theta,\dot{\theta}) = \left(\dot{r} + \tau^{-1}\sinh(r)\right)\sin\theta + \dot{\theta}\tanh(r)\cos\theta.$$

5.3.5.2 Directions of Vanishing Housekeeping Entropy Production

Having derived explicit expressions of $\mathcal{S}_{\mathrm{ex}}$ and $\mathcal{S}_{\mathrm{hk}}$, Eqs. (5.36) and (5.37), it is illuminating to consider points and trajectories on which $\mathcal{S}_{\mathrm{hk}}$ vanishes. Clearly this happens when $\theta(t) = 0$, $\theta(t) = \pi$, or $r(t) = 0$. This should come as no surprise since these parameters render the covariance $\Sigma(t)$ diagonal. However, this is not the only trajectory for which $\mathcal{S}_{\mathrm{hk}} = 0$.

Housekeeping entropy vanishes if and only if $\sigma(r,\dot{r},\theta,\dot{\theta}) = 0$. Separating the variables r and θ concludes

$$\frac{\dot{\theta}}{\tan\theta} = -\tau^{-1}\cosh(r) - \frac{\dot{r}}{\tanh(r)}.$$

Setting both sides equal to an arbitrary function of time $u(t)$ leads to

$$\dot{\theta} = \tan(\theta)u(t) \tag{5.38a}$$

$$\dot{r} = -\tau^{-1}\sinh(r) - \tanh(r)u(t), \tag{5.38b}$$

for any choice of function $u(t)$. Thus, trajectories corresponding to vanishing housekeeping entropy $\mathcal{S}_{\mathrm{hk}}$ are given by solutions to the system of Eqs. (5.38).

For example, the choice $u(t) = \frac{\kappa}{\tan(\theta_0+\kappa t)}$, with $\theta_0 = \theta(0)$ and $\kappa \geq 0$, reduces (5.38a) to $\dot{\theta} = \kappa$, while (5.38b) becomes

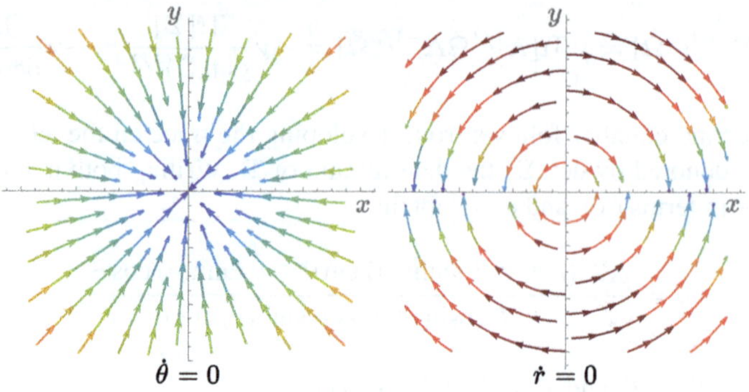

Fig. 5.2 Trajectories of vanishing $\mathcal{S}_{\text{hk}}$ in the (x, y) plane, where $x = r\cos(\theta)$ and $y = r\sin(\theta)$, and $\tau = 1$ in (5.38b). Left subplot: $u(t)$ is chosen such that $\dot{\theta} = 0$. Right subplot: $u(t)$ is chosen such that $\dot{r} = 0$. Image from [2]

$$\dot{r} = -\tau^{-1}\sinh(r) - \frac{\kappa}{\tan(\theta_0 + \kappa t)}\tanh(r) \quad r(0) = r_0. \tag{5.39}$$

This equation has a unique solution as long as $\theta_0 + \kappa t \in (0, \pi) \bmod \pi$. If we fix $\dot{\theta} = 0$, this is, $u(t) = 0$, then $\dot{r} = -\tau^{-1}\sinh(r)$ is always negative (but for when $r = 0$). Therefore, these trajectories always point towards the equilibrium point at $r = 0$. Similarly, one can choose $u(t)$ to keep $\dot{r}$ constant. If $\dot{r}$ is kept at 0, then $\dot{\theta} = -\tau^{-1}\cosh(r)\tan(\theta)$. These trajectories point towards either the $\theta = 0$ or $\theta = \pi$ equilibrium points depending on the initial state. The two sets of trajectories (with $\dot{\theta} = 0$ and $\dot{r} = 0$) are displayed in Fig. 5.2 for different initial conditions.

Thus, interestingly, from any initial state one can find trajectories with vanishing housekeeping entropy production that steer the system to any of the possible equilibrium states ($\theta \in \{0, \pi\}$, or $r = 0$). On the other hand, steering between states (not necessarily equilibrium states) while maintaining $\mathcal{S}_{\text{hk}} = 0$ impinges upon the controllability[4] of the control affine system (5.38) viewing $u(t)$ as a control input. It is clear that both $\theta = \{0, \pi\}$ and $r = 0$ constitute obstructions to controllability of (5.38), since the right-hand sides of (5.38a) and (5.38b) vanish, respectively. Moreover, from any state (r, θ), with $\theta \notin \{0, \pi\}$ and $r \neq 0$, a flow that maintains $\mathcal{S}_{\text{hk}} = 0$ can be selected within a cone of π radians. Specifically, accessible directions $\tan^{-1}(r\dot{\theta}/\dot{r})$ from (r, θ) are within the interval $[\alpha, \alpha + \pi]$, for

$$\alpha = \tan^{-1}\left(-\frac{r\tan(\theta)}{\tanh(r)}\right).$$

In light of (5.38), directions and therefore trajectories with vanishing housekeeping entropy production can not have arbitrarily small velocity fields; this can

[4] By controllability we refer to the ability to use the control to steer the system between any two states in finite time.

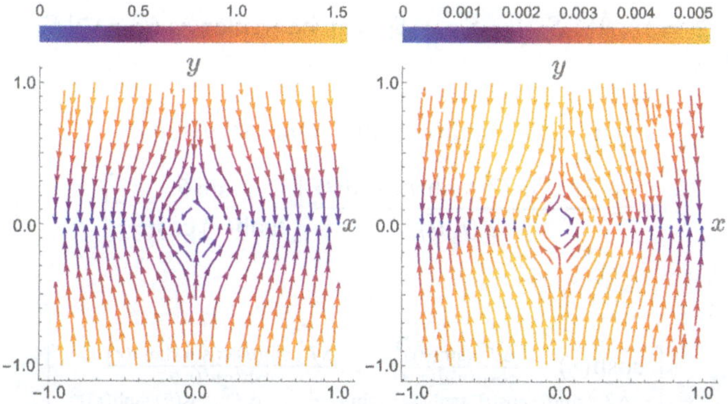

Fig. 5.3 Left: Trajectories that minimize excess entropy production while vanishing housekeeping entropy production. Right: Trajectories that minimize entropy production rate $\dot{\mathcal{S}}_{\mathrm{tot}}$. All trajectories are displayed in coordinates $x = r\cos(\theta)$, $y = r\sin(\theta)$ for the following choice of parameters: $\tau = 1 = \bar{T}$ and $\Delta T = 0.1$. Image from [2]

be traced to the fact that to eliminate circulation, the dynamical component in the decomposition (5.15) needs to cancel the steady-state component. This unavoidably leads to positive excess entropy production $\mathcal{S}_{\mathrm{ex}}$.

While maintaining $\dot{\mathcal{S}}_{\mathrm{hk}} = 0$, we may seek tangent directions $(\dot{r}, \dot{\theta})$ that also minimize excess entropy (in analogy to (5.19)), i.e.,

$$\min_u \, k_B \tau \int_0^{t_f} \left(\cosh(r)\dot{r}^2 + \sinh(r)\tanh(r)\dot{\theta}^2 \right) dt,$$

with $\dot{r}$ and $\dot{\theta}$ as in (5.38). This leads to the optimal choice of $u(t)$,

$$u^*(t) = -\tau^{-1} \frac{\cosh(r)}{1 + (\tan\theta)^2}.$$

Corresponding trajectories are displayed in the left subplot of Fig. 5.3, for a choice of parameters, and converge to a $\theta \in \{0, \pi\}$ equilibrium state.

5.3.5.3 Direction of Least Entropy Production Rate

Let us now characterize the direction of minimal entropy production *rate*.

From (5.36) and (5.37), the entropy rates for $\dot{\mathcal{S}}_{\mathrm{ex}}$ and $\dot{\mathcal{S}}_{\mathrm{hk}}$ are

$$\dot{\mathcal{S}}_{\mathrm{ex}} = k_B \tau \left(\cosh(r)\dot{r}^2 + \sinh(r)\tanh(r)\dot{\theta}^2 \right),$$

$$\dot{\mathcal{S}}_{\mathrm{hk}} = k_B \tau \frac{\Delta T^2}{\cosh(r)} \left(\frac{\sigma(r, \dot{r}, \theta, \dot{\theta})}{\bar{T} + \Delta T \tanh(r)\cos\theta} \right)^2.$$

Then, letting $\xi = [\dot{r}, \dot{\theta}]'$, $\dot{\mathcal{S}}_{\text{tot}} = \dot{\mathcal{S}}_{\text{ex}} + \dot{\mathcal{S}}_{\text{hk}}$ can be written as $(k_B\tau)^{-1}\dot{\mathcal{S}}_{\text{tot}} = \xi' A \xi + b'\xi + c$, where

$$c = \tau^{-2}\Delta T^2 \tanh(r)\sinh(r)(\sin\theta)^2/d(r,\theta),$$

$$b = \frac{2\tau^{-1}\Delta T^2 \tanh(r)\sin(\theta)}{d(r,\theta)}\begin{bmatrix}\sin\theta\\ \tanh(r)\cos(\theta)\end{bmatrix}$$

and

$$A = \begin{bmatrix}\cosh(r) + \frac{\Delta T^2(\sin(\theta))^2}{\cosh(r)d(r,\theta)} & \frac{\Delta T^2\sin(\theta)\cos(\theta)\tanh(r)}{\cosh(r)d(r,\theta)}\\ \frac{\Delta T^2\sin(\theta)\cos(\theta)\tanh(r)}{\cosh(r)d(r,\theta)} & \frac{(\sinh r)^2}{\cosh(r)} + \frac{\Delta T^2(\cos(\theta)\tanh(r))^2}{\cosh(r)d(r,\theta)}\end{bmatrix},$$

with $d(r,\theta) = (\bar{T} + \Delta T\tanh(r)\cos\theta)^2$. Since A is positive definite for all r and θ, we obtain that $\xi = -A^{-1}b'/2$ minimizes $\dot{\mathcal{S}}_{\text{tot}}$ over all possible ξ. This is, for any given $(r,\theta) \in [0,\infty) \times [0, 2\pi)$, the directions $(\dot{r}, \dot{\theta})$ that minimize the entropy production rate $\dot{\mathcal{S}}_{\text{tot}}$ are

$$\dot{r} = -\tau^{-1}h(r,\theta)\sinh(r)\sin(\theta), \tag{5.40a}$$

$$\dot{\theta} = -\tau^{-1}h(r,\theta)\cosh(r)\cos(\theta), \tag{5.40b}$$

where

$$h(r,\theta) = \frac{(\tanh(r))^2\sin(\theta)}{(\frac{\bar{T}}{\Delta T} + \cos(\theta)\tanh(r))^2(\sinh(r))^2 + (\tanh(r))^2}.$$

Optimal trajectories for different initial conditions are drawn in the right subplot in Fig. 5.3. It is worth noting that these streamlines are similar *in form* to those to the left, which correspond to trajectories that minimize excess entropy production while vanishing housekeeping entropy production.

5.3.5.4 Minimal Entropy Production Close to Equilibrium

We now return to the problem of minimizing total entropy production over a path that joins two (possibly non-equilibrium) end-points in finite time. To gain insight into the nature of optimal trajectories, we consider staying close to the equilibrium that corresponds to $r = 0$.

To this end, we let $r(t) = \epsilon\hat{r}(t)$, $\hat{x} = \hat{r}\cos\theta$ and $\hat{y} = \hat{r}\sin\theta$, and expand the expression for $\mathcal{S}_{\text{tot}}$ in terms of $\epsilon > 0$, assumed small, to obtain

$$\mathcal{S}_{\text{tot}} = \epsilon^2 k_B \tau \int_0^{t_f} \left(\dot{\hat{r}}^2 + \hat{r}^2\dot{\theta}^2 + \left(\frac{\Delta T}{\bar{T}}\right)^2 \left((\dot{\hat{r}} + \tau^{-1}\hat{r})\sin\theta + \dot{\theta}\cos\theta\hat{r}\right)^2\right) dt + O(\epsilon^3)$$

$$= \underbrace{\epsilon^2 k_B \tau \int_0^{t_f} \left(\dot{\hat{x}}^2 + \dot{\hat{y}}^2 + \left(\frac{\Delta T}{\bar{T}}\right)^2 (\dot{\hat{y}} + \tau^{-1}\hat{y})^2\right) dt}_{\mathcal{S}_\epsilon(\hat{x},\hat{y})} + O(\epsilon^3),$$

where we have defined the entropy production up to second-order in ϵ, $\mathcal{S}_\epsilon(\hat{x}, \hat{y})$. Consider the problem of minimizing $\mathcal{S}_\epsilon$, i.e.

$$(\hat{x}^*, \hat{y}^*) = \operatorname{argmin}\{\mathcal{S}_\epsilon(\hat{x}, \hat{y}) \mid \hat{x}(0) = \hat{x}_0,\ \hat{y}(0) = \hat{y}_0,\ \hat{x}(t_f) = \hat{x}_{t_f},\ \hat{y}(t_f) = \hat{y}_{t_f}\}. \tag{5.41}$$

The corresponding Euler-Lagrange equations take the form

$$\ddot{\hat{x}} = 0, \quad \hat{x}(0) = \hat{x}_0,\ \hat{x}(t_f) = \hat{x}_{t_f}, \tag{5.42}$$

$$\ddot{\hat{y}} = \frac{1}{\hat{\tau}^2}\hat{y}, \quad \hat{y}(0) = \hat{y}_0,\ \hat{y}(t_f) = \hat{y}_{t_f}, \tag{5.43}$$

where $\hat{\tau} = \tau\left(1 + \left(\frac{\bar{T}}{\Delta T}\right)^2\right)^{1/2}$. Solving these equations and imposing the boundary conditions we obtain

$$\hat{x}^*(t) = \hat{x}_0 + \frac{t}{t_f}(\hat{x}_{t_f} - \hat{x}_0), \tag{5.44a}$$

$$\hat{y}^*(t) = c_+ e^{t/\hat{\tau}} + c_- e^{-t/\hat{\tau}}. \tag{5.44b}$$

with

$$c_\pm = \frac{\hat{y}_0 - \hat{y}_{t_f} e^{\pm t_f/\hat{\tau}}}{1 - e^{\pm 2t_f/\tau}}.$$

It is also insightful to consider fixing the starting state at $(\hat{x}_0, \hat{y}_0)$, near the equilibrium at $r = 0$ as before, and consider the trajectory departing from this state with terminal time $t_f \to \infty$ and the final state unconstrained. The optimal solution is then given by $\hat{x}(t) = \hat{x}_0$ and $\hat{y}(t) = \hat{y}_0 e^{-t/\hat{\tau}}$, as any other $\hat{y}(t)$ would lead to infinite entropy production. Therefore, as long as r is small enough, trajectories minimizing entropy production over an unbounded time interval end up at one of the $\theta \in \{0, \pi\}$ equilibrium states (as opposed to the one corresponding to $r = 0$).

It is interesting to observe that trajectories minimizing total entropy production over an infinite time window, for r small enough, share some resemblance with those of vanishing housekeeping entropy production while minimizing excess entropy rate, as well as those minimizing total entropy production rate, portrayed in the left and right subplots of Fig. 5.3, respectively. Indeed, they approach equilibrium for $\theta \in \{0, \pi\}$ almost vertically and have τ as a natural time constant.

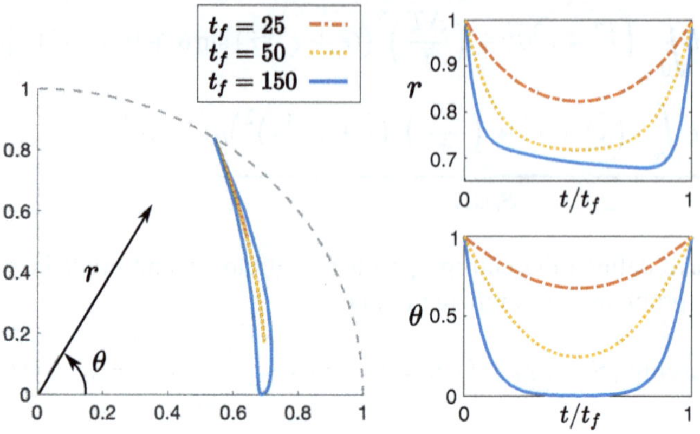

Fig. 5.4 Entropy minimizing cycles starting and ending at $(r, \theta) = (1, 1)$, with final times $t_f = 25$, 50 and 150 seconds, respectively. The rest of the parameters are set to 1, but for $\Delta T = 0.1$. Image from [2]

5.3.5.5 Entropy Minimizing Cycles

In this last section, we turn to cycles that generate minimal entropy production. The expressions we obtained did not allow for closed-form solutions, and hence we have resorted to numerically computing optimal trajectories. Valuable insights are gained in that we observe a natural tendency of trajectories to gravitate towards an equilibrium state, as much as time allows, before returning to their starting point. In fact, we identify a time constant that dictates the speed at which equilibrium is approached. This is in stark contrast to the *isotropic* case with a single heat bath, in which optimal trajectories are trivially constant.

We selected as the starting and ending point $(r_0, \theta_0) = (1, 1)$, and computed closed trajectories, portrayed in Fig. 5.4, of different periods. The natural tendency is to gravitate towards an equilibrium state where $\mathcal{S}_{\text{hk}} = 0$, interestingly, one that corresponds to $\theta = 0$. We have seen the same tendency in the analysis of small r, where entropy minimizing trajectories over an infinite time window converge to an equilibrium at $y = 0 \Leftrightarrow \theta = 0$; this is apparent in Fig. 5.5.

Figure 5.5 compares the analytic solution when minimizing entropy production for r vanishingly small (5.41), to the numerical solution in Fig. 5.4 that is computed for r large. At the outset, there is no apparent reason why these should compare. Yet, it is observed that the θ component of trajectories, *whether r is small or not*, follows a similar path. To see this, we drew in Fig. 5.5 the θ component of the analytic solution (5.44) as a function of time (continuous curve), this is $\theta(t) = \tan^{-1}\left(\frac{\hat{y}^*(t)}{\hat{x}^*(t)}\right)$ with $\hat{x}^*$ and $\hat{y}^*$ as in (5.44). A very good agreement with the numerical solution is readily seen (marked with $\times$'s). The r-components cannot be compared and thus respective plots are omitted.

An additional reason for focusing on the θ component of trajectories is that this component reveals an apparent natural time constant $\hat{\tau}$ that dictates an optimal

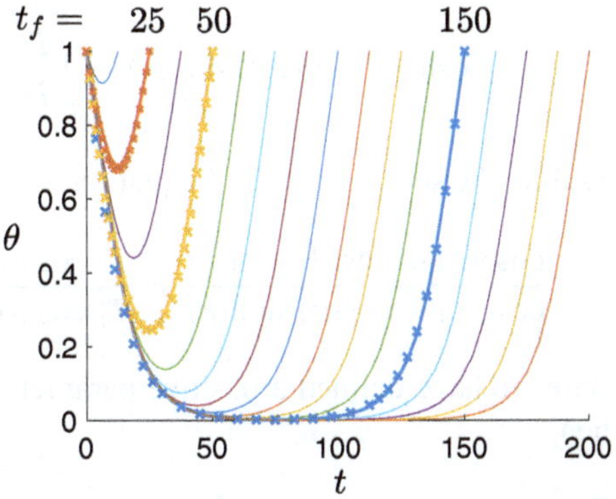

Fig. 5.5 Solid lines represent θ component of entropy minimizing cycles in the limit of vanishing r for different final times t_f. Crosses denote the θ component of entropy minimizing cycles obtained numerically (not requiring r small). The rest of the parameters are set to 1, but for $\Delta T = 0.1$. Image from [2]

balance between $\mathcal{S}_{\text{hk}}$ and $\mathcal{S}_{\text{ex}}$, as the trajectories approach a suitable point of equilibrium at $\theta = 0$. Indeed, when the period is sufficiently large, trajectories spend any "extra" time allocated near the equilibrium, as seen in Fig. 5.5. Interestingly, within the time window that the trajectory stays near equilibrium (near $\theta = 0$), assuming t_f is sufficiently large, r linearly decreases as seen in Fig. 5.4. Apparently, this helps in reducing the housekeeping cost during the return leg of the trajectory back to $(r, \theta) = 1$.

We remark that entropy minimizing cycles are work consuming [24, 25]. Specifically, to steer the thermodynamic state along these closed trajectories, the controlling input (actuating potential) needs to supply work to the system. This can be seen from Sect. 4.1.3, and in particular Fig. 4.1. Therein we showed that clockwise traversed closed trajectories in the upper-half plane of the (r, θ) space indicate work consumption. Indeed, entropy minimizing trajectories are clockwise traversed (as can be seen from the asymmetry in the r subplot of Fig. 5.4).

However, note that the (r, θ) space in the present Section does not correspond to the one introduced in Sect. 4.1.3, since we are now parameterizing $\widetilde{\Sigma} = T^{-1/2}\Sigma T^{-1/2}$. Let us momentarily refer to the parameters of Sect. 4.1.3 as r_0 and θ_0. To ensure that entropy minimizing trajectories are work consuming, we need to ensure that clockwise traversed paths in the upper-half (r, θ) plane, remain clockwise traversed paths in the upper-half (r_0, θ_0) plane.

To show this, it is sufficient to ensure the positivity of the Jacobian of the transformation from $(r, \theta) \to (r_0, \theta_0)$, this is, the transformation that satisfies $\widetilde{\Sigma}(r, \theta) = T^{-1/2}\Sigma(r_0, \theta_0)T^{-1/2}$. After taking the products of the matrices on both sides, we obtain

$$\sin(\theta)\sinh(r) = \sin(\theta_0)\sinh(r_0) \tag{5.45}$$

$$\cosh(r) + \cos(\theta)\sinh(r) = \sqrt{\frac{T_2}{T_1}}(\cosh(r_0) + \cos(\theta_0)\sinh(r_0))$$

$$\cosh(r) - \cos(\theta)\sinh(r) = \sqrt{\frac{T_1}{T_2}}(\cosh(r_0) - \cos(\theta_0)\sinh(r_0)).$$

Dividing the second and third equations we obtain

$$\frac{\cosh(r) + \cos(\theta)\sinh(r)}{\cosh(r) - \cos(\theta)\sinh(r)} = \frac{T_2}{T_1}\frac{\cosh(r_0) + \cos(\theta_0)\sinh(r_0)}{\cosh(r_0) - \cos(\theta_0)\sinh(r_0)} =: \alpha(r, \theta),$$

where we have defined a positive parameter α (positive since $\cosh(x) \geq \sinh(x)$). Then,

$$\frac{\cosh(r) + \cos(\theta)\sinh(r)}{\cosh(r) - \cos(\theta)\sinh(r)} = \alpha \implies \frac{\cos(\theta)\sinh(r)}{\cosh(r)} = \frac{\alpha - 1}{\alpha + 1} =: \kappa(\alpha).$$

Let us also define the right-hand-side of (5.45) as $\beta := \sin(\theta_0)\sinh(r_0)$, then

$$\sin(\theta)\sinh(r) = \beta(r, \theta).$$

The sign of the Jacobian of the transformation from (r, θ) to (r_0, θ_0) is positive if the Jacobian of the transformations $(r, \theta) \to (\kappa, \beta)$ and $(\alpha, \beta) \to (\kappa, \beta)$ are positive. Indeed,

$$\det\begin{bmatrix} \partial_r\kappa & \partial_\theta\kappa \\ \partial_r\beta & \partial_\theta\beta \end{bmatrix} = \det\begin{bmatrix} \frac{\cos(\theta)}{\cosh(r)^2} & -\sin(\theta)\tanh(r) \\ \sin(\theta)\cosh(r) & \cos(\theta)\sinh(r) \end{bmatrix}$$

$$= \frac{\cos(\theta)^2\sinh(r)}{\cosh(r)^2} + \sin(\theta)^2\sinh(r) > 0$$

and

$$\det\begin{bmatrix} \partial_\alpha\kappa & \partial_\beta\kappa \\ \partial_\alpha\beta & \partial_\beta\beta \end{bmatrix} = \det\begin{bmatrix} \frac{2}{(\alpha+1)^2} & 0 \\ 0 & 1 \end{bmatrix} > 0.$$

Hence, we may conclude that entropy minimizing cycles are traversed clockwise in the upper-(r_0, θ_0)-plane, and are thus work consuming.

Finally, we would like to underscore the importance of controlling the thermodynamic system for positive work output, balancing the work produced with total entropy generated. This optimal control problem remains open at present. Indeed, in Chap. 4 we focused on steering the anisotropic thermodynamic system over a cycle that maximizes work extraction. In light of the present results, it is worth considering maximizing work output subject to a bound on the entropy production over a cycle. Such a bound is natural, especially in biological engines, where sources that sustain chemical gradients are taxed by their entropy production. Optimal control of thermodynamic systems for the purpose of trading off entropy for work may prove essential in the understanding of biological mechanisms that power living systems.

References

1. Movilla Miangolarra, O., Taghvaei, A., Georgiou, T.T.: Minimal entropy production in anisotropic temperature fields. In: 2023 American Control Conference (ACC), pp. 616–621. IEEE (2023)
2. Movilla Miangolarra, O., Taghvaei, A., Georgiou T.T.: Minimal entropy production in the presence of anisotropic fluctuations. In: IEEE Transactions on Automatic Control (2025)
3. Aurell, E., Mejía-Monasterio, C., Muratore-Ginanneschi, P.: Optimal protocols and optimal transport in stochastic thermodynamics. IEEE Transactions on Automatic Control. Phys. Rev. Lett. **106**, 1558–2523 (2011)
4. Dechant, A., Sakurai, Y.: Thermodynamic interpretation of Wasserstein distance. Preprint (2019). arXiv:1912.08405
5. Chen, Y., Georgiou, T., Tannenbaum, A.: Stochastic control and non-equilibrium thermodynamics: fundamental limits. IEEE Trans. Autom. Control **65**(1), 252–262 (2020)
6. Van Vu, T., Saito, K.: Thermodynamic unification of optimal transport: Thermodynamic uncertainty relation, minimum dissipation, and thermodynamic speed limits. Phys. Rev. X **13**, 011013 (2023)
7. Battle, C., Broedersz, C.P., Fakhri, N., Geyer, V.F., Howard, J., Schmidt, C.F., MacKintosh, F.C.: Broken detailed balance at mesoscopic scales in active biological systems. Science (2016)
8. Gnesotto, F.S., Mura, F., Gladrow, J., Broedersz, C.P.: Broken detailed balance and non-equilibrium dynamics in living systems: a review. Reports on Progress in Physics, IOP publishing (2018)
9. Ambrosio, L., Gigli, N., Savaré, G.: Gradient flows: in metric spaces and in the space of probability measures. Springer Science & Business Media, Cham (2005)
10. Morse, P.M., Feshbach, H.: Methods of Theoretical Physics. International Series in Pure and Applied Physics. McGraw-Hill book company, inc. (1953)
11. Chorin, A.J., Marsden, J.E., Marsden, J.E.: A Mathematical Introduction to Fluid Mechanics, vol. 3. Springer, Berlin (1990)
12. Villani, C.: Topics in Optimal Transportation. American Mathematical Society, Providence (2003)
13. Ciliberto, S., Imparato, A., Naert, A., Tanase, M.: Statistical properties of the energy exchanged between two heat baths coupled by thermal fluctuations. J. Stat. Mech. Theory Exp. **2013**(12), P12014 (2013)
14. Ciliberto, S., Imparato, A., Naert, A., Tanase, M.: Heat flux and entropy produced by thermal fluctuations. Phys. Rev. Lett. **110**, 180601 (2013)
15. Chiang, K.-H., Lee, C.-L., Lai, P.-Y., Chen, Y.-F.: Electrical autonomous Brownian gyrator. Phys. Rev. E **96**, 032123 (2017)
16. Argun, A., Soni, J., Dabelow, L., Bo, S., Pesce, G., Eichhorn, R., Volpe, G.: Experimental realization of a minimal microscopic heat engine. Phys. Rev. E **96**, 052106 (2017)
17. Nakazato, M., Ito, S.: Geometrical aspects of entropy production in stochastic thermodynamics based on wasserstein distance. Phys. Rev. Res. (2021)
18. Dechant, A., Sasa, S.-I., Ito, S.: Geometric decomposition of entropy production in out-of-equilibrium systems. Phys. Rev. Res. (2022)
19. Ito, S.: Geometric thermodynamics for the fokker–planck equation: stochastic thermodynamic links between information geometry and optimal transport. Inf. Geom. 1–42 (2023)
20. Dechant, A., Sasa, S., Ito, S.: Geometric decomposition of entropy production into excess, housekeeping, and coupling parts. Phys. Rev. E **106**, 024125 (2022)

21. Yoshimura, K., Kolchinsky, A., Dechant, A., Ito, S.: Housekeeping and excess entropy production for general nonlinear dynamics. Phys. Rev. Res. **5**, 013017 (2023)
22. Van Handel, R.: Stochastic calculus, filtering, and stochastic control. Course Notes (2007). http://www.princeton.edu/rvan/acm217/ACM217.pdf14
23. Laugesen, R.S., Mehta, P.G., Meyn, S.P., Raginsky, M.: Poisson's equation in nonlinear filtering. SIAM J. Control Optim. (2015)
24. Movilla Miangolarra, O., Taghvaei, A., Fu, R., Chen, Y., Georgiou, T.T.: Energy harvesting from anisotropic fluctuations. Phys. Rev. E (2021)
25. Movilla Miangolarra, O., Taghvaei, A., Chen, Y., Georgiou, T.T.: Geometry of finite-time thermodynamic cycles with anisotropic thermal fluctuations. IEEE Control Syst. Lett. **6**, 3409–3414 (2022)

Chapter 6
Application: Thermodynamic Engine Powered by Anisotropic Fluctuations

Inspired by nature's ability to harvest energy from fluctuations and anisotropic chemical concentrations together with varying electrochemical potentials [1, 2], we introduce an "engine concept" based on the model of a Brownian gyrator that *autonomously* extracts energy from thermal anisotropy through a periodically varying potential. To do so, we explore the coupling between (fast) thermal fluctuations and a (slow) mechanical component in a way that allows generation of mechanical power.

Specifically, we consider the electrical embodiment of the Brownian gyrator [3–5], which consists of two resistors and three capacitors. Johnson-Nyquist fluctuating currents at the two resistors, due to uneven ambient temperatures, allow for the potential to generate torque. We postulate variable capacitors with a moveable dielectric material. Forces on the dielectric material are exerted by the fluctuating currents as well as by a coupling to a flywheel. This mechanical component provides inertia and dissipation that absorbs generated power.

Moreover, we show that one can dispense with this inertia, often uncharacteristic of biological systems, by coupling multiple subsystems [6]. We provide a detailed analysis as a proof-of-concept for the feasibility (and stability) of this power generating mechanism, and explain the mechanics responsible for energy transfer between the fluctuating currents and the rotational subcomponent of the engine.

Proposals for heat engines that are powered by thermal excitations date back to Maxwell's demon [7, 8] and the Feynman ratchet [9]. Experiments to validate relevant thought experiments have been reported recently [10–15]. However, these experimental demonstrations are mostly based on manipulating particles in a non-autonomous manner, via optical traps and externally specified cyclic control protocols [16, 17], similar to the ones we have presented up until now. This nano-manipulation requires considerable energy that far exceeds work that is being produced. The point of this chapter is to present an analysis for the coupling between the system responsible for the anisotropy of temperature with a slower mechanical component that renders the operation of the engine autonomous [18].

O. Movilla Miangolarra, *Stochastic Thermodynamic Treatment of Thermal Anisotropy*, Springer Theses, https://doi.org/10.1007/978-3-031-68066-3_6

6.1 Electrical Embodiment of the Brownian Gyrator

We consider the electrical embodiment of a Brownian gyrator (the same as in Sect. 3.4.2, repeated here for convenience) that consists of three capacitors and two resistors as shown in Fig. 6.1. The resistors are in contact with heat baths of different temperatures. The state of this electrical-thermal system is comprised of the charges at two of the capacitors. Specifically, let $q_1(t)$ and $q_2(t)$ denote the charges at capacitances $C_1(t)$ and $C_2(t)$ (that are time-varying), and set

$$q_t = \begin{bmatrix} q_1(t) \\ q_2(t) \end{bmatrix}, \quad R = \begin{bmatrix} R_1 & 0 \\ 0 & R_2 \end{bmatrix}, \quad \text{and } C(t) = \begin{bmatrix} C_1(t) + C_c(t) & -C_c(t) \\ -C_c(t) & C_2(t) + C_c(t) \end{bmatrix},$$

for the charge vector, and the resistance and capacitance matrices, respectively. Basic physics dictates that the capacitance matrix $C(t)$ is positive definite at all times t. The dynamics of this system follow a two-dimensional stochastic differential equation,

$$dq_t = -R^{-1}C^{-1}(t)q_t dt + R^{-1}DdB_t, \tag{6.1}$$

with

$$D = \begin{bmatrix} \sqrt{2k_B R_1 T_1} & 0 \\ 0 & \sqrt{2k_B R_2 T_2} \end{bmatrix},$$

and T_1 and T_2 the temperature to which the corresponding resistors are subjected to.

As already mentioned in Sect. 3.4.2, the state Eq. (6.1) becomes a two-dimensional overdamped Langevin equation in the absence of non-conservative forces,

$$dq_t = -R^{-1}\nabla U(t, q_t)dt + R^{-1}DdB_t,$$

where $U(t, q) = \frac{1}{2}q'C^{-1}(t)q$ is the potential energy of the system of capacitances and ∇ denotes the gradient with respect to q. The state of the stochastic system is the probability distribution of charges, denoted by $\rho(t, q)$. It satisfies the Fokker-Planck equation

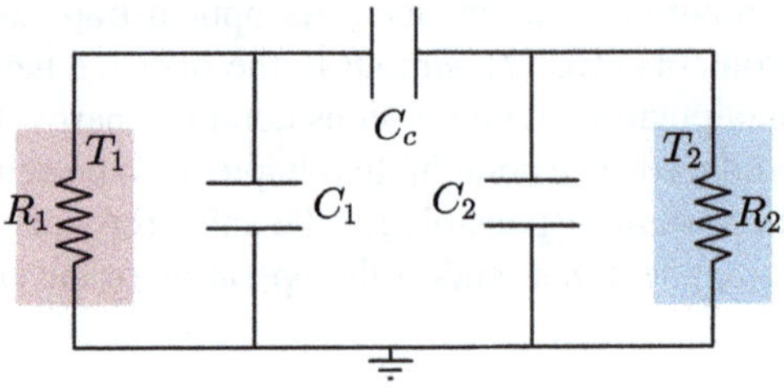

Fig. 6.1 Electrical embodiment of the Brownian gyrator [5] (image from [18])

$$\frac{\partial \rho}{\partial t} + \nabla \cdot J = 0,$$

with probability current

$$J = -R^{-1}\left[\nabla U + \frac{1}{2}DD'R^{-1}\nabla \log(\rho)\right]\rho.$$

If the initial state $\rho(t, q)$ is Gaussian with mean 0 and covariance Σ_0, then, under time-varying quadratic potential (as in here), $\rho(t, q)$ remains Gaussian for all times t, with mean 0 and covariance $\Sigma(t)$ that satisfies the Lyapunov equation (see Sect. 3.4 for the derivation)

$$\dot{\Sigma}(t) = -R^{-1}C^{-1}(t)\Sigma(t) - \Sigma(t)C^{-1}(t)R^{-1} + R^{-1}DD'R^{-1}. \tag{6.2}$$

The coupling in (6.1) with the two heat baths allows transference of heat between the two, as well as exchange of energy with the environment through coupling with the time-varying potential U. Indeed, the total energy in the system (averaged over realizations) is

$$\mathcal{E} = \int U\rho\, dq = \frac{1}{2}\,\mathrm{Tr}[C^{-1}(t)\Sigma(t)],$$

while the power delivered to the system via changes in the potential can be written as

$$đ\mathcal{W} = \mathbb{E}[đW] = \int \partial_t U\rho\, dq dt = \frac{1}{2}\,\mathrm{Tr}[\dot{C}^{-1}(t)\Sigma(t)]dt,$$

where

$$đW = \partial_t U dt$$

denotes the work rate along a single realization of the process.

In the sequel, we are interested in the case where the capacitances $C_k(t)$ ($k \in \{c, 1, 2\}$) vary with time so as to allow extracting thermodynamic work out of the system. In Chap. 4, we quantified trade-offs between dissipation and work that can be extracted via externally and periodically varying parameters [19, 20]. Herein, we pursue an alternative route where the relevant parameters (capacitances) are a function of added degrees of freedom, introduced via coupling of the components, specifically the dielectric material in the capacitors, with an inertial wheel. This allows the autonomous function of the Brownian gyrator as a genuine autonomous thermodynamic engine.

6.2 Engine Concept

Let us consider the presence of dielectric padding in the three capacitors, which can vary in position through a mechanical coupling to a rotating wheel, as shown with a schematic in Fig. 6.2. In this way, the angular position θ_t of the (inertial) wheel forces the dielectric material in and out of the respective capacitors. This mechanical coupling renders the capacitance matrix variable with time, being function of the dynamic variable θ_t.

We select a geometry of the linkages actuating the dielectric material that gives the capacitance matrix as a function of θ in the form

$$C(\theta) = C_0 \begin{bmatrix} 2+\beta g_1(\theta) & -1-\beta\cos(\theta) \\ -1-\beta\cos(\theta) & 2+\beta g_2(\theta) \end{bmatrix},$$

where $g_1(\theta) = \cos(\theta+\phi_1)+\cos(\theta)$, $g_2(\theta) = \cos(\theta+\phi_2)+\cos(\theta)$ and $0 < \beta < 1$. The specific form follows if we assume that i) the linkages are long enough so that the capacitances, to a good approximation, vary sinusoidally with the angular position θ of the wheel, and ii) that the links are attached to suitable positions to account for the phase differences ϕ_1, ϕ_2.

The purpose of the mechanical coupling is to transfer the torque generated by the thermal fluctuations at the capacitors to the inertial wheel. The inertial wheel then averages out the fluctuations and provides the needed phase difference (reflected in the parameters ϕ_1, ϕ_2) between the elements of the engine so as to sustain its continuous operation.

The model for the coupled system of the (electrical) Brownian gyrator, the inertial subsystem with linkages shown in Fig. 6.2, and the external torque $-\Gamma\omega_t$ to transfer work to the environment, is

$$dq_t = -R^{-1}C^{-1}(\theta_t)q_t dt + R^{-1}DdB_t, \tag{6.3a}$$

$$d\theta_t = \omega_t dt, \tag{6.3b}$$

$$I d\omega_t = -\frac{1}{2}q_t'\partial_\theta C^{-1}(\theta_t)q_t dt - \Gamma\omega_t dt. \tag{6.3c}$$

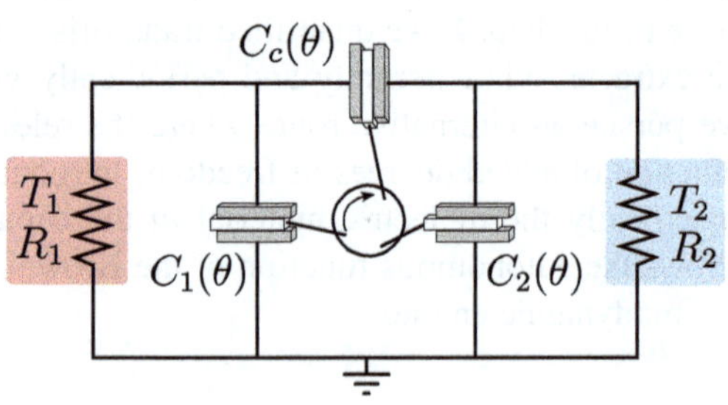

Fig. 6.2 Engine embodiment with actuated dielectric pads. The angular position θ of the wheel forces the dielectric material in and out of the capacitors, rendering the capacitance matrix a function of θ. Image extracted from [6]

In the above, the symbol I represents the inertia of the wheel, and Γ can be thought of as a friction coefficient in a process that helps extracting the work out of the engine. We will refer to the term $-\Gamma\omega_t$ as external dissipation, though it could just as well represent torque proportional to ω exchanged with an external subsystem.

Noticing that

$$đW = \partial_t U(t, q_t)dt = \frac{1}{2} q_t' \partial_\theta C^{-1}(\theta_t) q_t \omega_t dt,$$

we rewrite (6.3c),

$$I\omega_t d\omega_t = -đW - \Gamma\omega_t^2 dt,$$

and integrate over time from 0 to t to obtain

$$\frac{1}{2} I(\omega_t^2 - \omega_0^2) = -W_t - \Gamma \int_0^t \omega_s^2 ds,$$

where we have defined $W_t = \int_0^t đW$. Therefore, the change in the kinetic energy of the wheel equals the work produced by the engine $(-W_t)$, minus the energy transferred via the torque $\Gamma\omega_t$ to the environment (as friction or coupled to another system). Thus, it is intuitively clear that as long as the engine produces work, the parameter Γ can be adjusted to ensure that the wheel keeps rotating.

6.2.1 Analysis

We present an analysis that supports our claim and shows that, for a suitable set of parameters, the autonomous system generates positive work output over a cycle which sustains the rotational motion of the wheel and at the same time supplies torque to an external dissipative load.

We adopt the assumption that there is a significant time-scale separation between the electrical and mechanical subcomponents of the engine, in that $\tau_{\text{elec}} \ll \tau_{\text{mech}}$, where τ_{elec} and τ_{mech} are the time-scales governing the electric and mechanical subsystems, respectively. With this time-scale separation, at every time-instant, q_t can be viewed as a random vector that follows the stationary distribution of (6.3a) associated with θ_t. The distribution is Gaussian with zero mean and covariance matrix $\Sigma(\theta)$ that satisfies the algebraic Lyapunov equation

$$-R^{-1}C^{-1}(\theta)\Sigma(\theta) - \Sigma(\theta)C^{-1}(\theta)R^{-1} + R^{-1}DD'R^{-1} = 0. \tag{6.4}$$

Moreover, the correlation of q_t is localized over time due to its fast dynamics. Thus, the randomness it injects to (6.3c) averages out fast so that, effectively, the mechanical subsystem is driven by the covariance $\Sigma(\theta)$ of the charge vector q_t.

Hence, the dynamics of the mechanical components can be approximated by the deterministic dynamics

$$\dot{\theta} = \omega, \tag{6.5a}$$

$$I\dot{\omega} = -\frac{1}{2}\operatorname{Tr}[\partial_\theta C^{-1}(\theta)\Sigma(\theta)] - \Gamma\omega. \tag{6.5b}$$

The accuracy of the above approximation is positively correlated with the magnitude of the time-scale separation. Elaborating further, the assumption on time-scale separation allows us to write θ and ω as functions of the time average of $q_t q_t'$ over τ_{mech}, i.e., $\frac{1}{\tau_{\text{mech}}}\int_t^{t+\tau_{\text{mech}}} q_s q_s' ds$. Since the stochastic system (6.3a) is linear and the matrix $R^{-1}C^{-1}$ has positive eigenvalues, q_t will approach, relatively fast, a steady state Gaussian distribution with variance $\Sigma(\theta)$. Thereby, we can replace this time average with $\Sigma(\theta)$.

6.2.1.1 Phase Portrait

The system of Eq. (6.5) describes motion in a force field,

$$I\ddot{\theta} + \Gamma\dot{\theta} = F(\theta)$$

where $F(\theta) = -\operatorname{Tr}[\partial_\theta C^{-1}(\theta)\Sigma(\theta)]/2$. The dynamics are analogous to those of a forced pendulum [21], and while in our case F may not be exactly sinusoidal, qualitatively the response is quite similar.

A schematic of a phase plot, for a suitable choice of parameters, is shown in Fig. 6.3. In this, the red dotted curve highlights the isocline[1]

$$\omega = \frac{1}{\Gamma}F(\theta),$$

which is periodic in θ. The flow field above this curve (on the upper-half place) points "southeast," whereas below the curve, it points "northeast." Points of equilibrium exist at the intersection of the isocline with the $\omega = 0$ axis, i.e., at points $(\theta, \omega = 0)$ where $F(\theta) = 0$. Over a 2π-interval there are two such points of equilibrium. The first, labeled p_1 in Fig. 6.3, is a stable focus, whereas the second, labeled p_2, is a saddle point (unstable). Their pattern repeats with a period equal to 2π.

Only two types of trajectories are possible. Either the system settles to an oscillatory motion about a point of equilibrium, or it settles about a persistent periodic orbit where θ monotonically increases. Such a trajectory "weaves" with the isocline and is highlighted in the figure (Fig. 6.3) by a periodic solid line.

---

[1] The curve with zero acceleration.

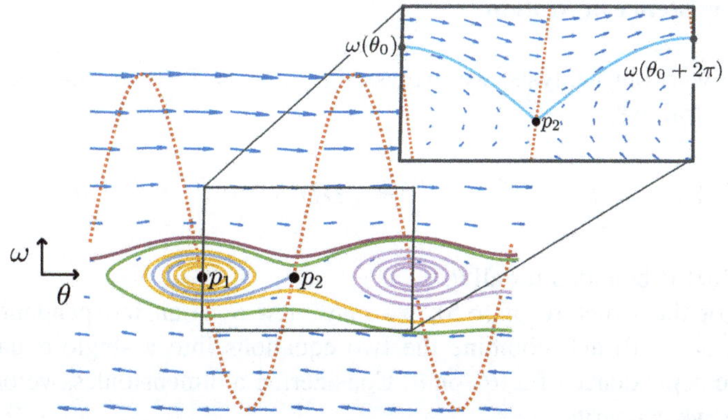

Fig. 6.3 Schematic phase portrait of (6.5). The red dotted line marks the isocline, while solid lines display different solutions in state space (θ, ω). The point p_1 and p_2 label stable and unstable equilibrium points, respectively. In the blow-up figure we have displayed a solution that crosses the neighborhood of the saddle point from above. The behavior of this solution helps establish the existence of the stable periodic orbit. This periodic orbit is seen in the figure to "weave" at about 90°-phase difference with the isocline. Image from [18]

The existence of a stable periodic orbit requires sufficient work production over a cycle. A sufficient condition that guarantees existence of such a stable periodic orbit can be deduced from the Poincare map P that relates the value of $\omega(\theta)$ along orbits corresponding to values of θ that differ by 2π:

$$P : \omega(\theta_0) \mapsto \omega(\theta_0 + 2\pi).$$

To this end, we consider a trajectory that "scrapes" past the unstable equilibrium at p_2 as shown in the blowup of Fig. 6.3. This trajectory can be numerically evaluated integrating forward and backward in time, starting from a neighborhood of p_2. The inequality $\omega(\theta_0 + 2\pi) \geq \omega(\theta_0)$ guarantees the existence of a periodic orbit.

In fact, if this periodic orbit exists, it must be unique, since there can only exist one periodic orbit that dissipates exactly the amount of work that is produced over a cycle. Specifically, work produced over a cycle is given by

$$W_{\text{cycle}} = -\frac{1}{2}\int_0^{2\pi} \text{Tr}[\partial_\theta C^{-1}(\theta)\Sigma(\theta)]d\theta,$$

independently of the initial conditions. On the other hand, dissipation over a cycle is given by $\Gamma \int_0^{2\pi} \omega(\theta)d\theta$, which is monotonic in the initial velocity $\omega(\theta_0)$. Therefore, if a periodic orbit exists, it is unique, since trajectories in the phase portrait can not cross, and thus only one solution curve can satisfy $\Gamma \int_0^{2\pi} \omega(\theta)d\theta = W_{\text{cycle}}$.

Our phase portrait analysis, hence, helps assess numerically the existence of such an orbit. In the following, we pursue an alternative route to ensure the existence of a periodic orbit for a family of parameters.

6.2.1.2 Persistence of Motion

To continue with our analysis, we express the solution $\Sigma(\theta)$ of (6.4) explicitly in terms of θ as follows,

$$\Sigma(\theta) = \int_0^\infty e^{-R^{-1}C^{-1}(\theta)\tau} R^{-1}DD'R^{-1} e^{-R^{-1}C^{-1}(\theta)\tau} d\tau.$$

Note that $\Sigma(\theta)$ is bounded for all θ.

In light of the structure of (6.5), we may view θ as an independent variable (as long as $\omega > 0$) and combine the two equations into a single equation that specifies the dependence of $\omega(\theta)$ on θ. Considering a dimensionless velocity $\Omega = \Gamma/(k_B T_1)\omega$, we can write

$$\frac{d\Omega}{d\theta}(\theta) = \epsilon f(\theta, \Omega), \tag{6.6}$$

where $\epsilon = \Gamma^2/(I k_B T_1)$, and

$$f(\theta, \Omega) = -\frac{1}{2k_B T_1}\frac{1}{\Omega}\text{Tr}[\partial_\theta C^{-1}(\theta)\Sigma(\theta)] - 1$$

is a 2π-periodic function of θ. Note that $f(\theta, \Omega)$ is continuous and bounded, and so are its derivatives with respect to Ω up to second order on $(\theta, \Omega) \in [0, \infty) \times [\Omega_{\text{min}}, \Omega_{\text{max}}]$ with $\Omega_{\text{max}} > \Omega_{\text{min}} > 0$.

We can infer stability from the averaged system as long as ϵ is small enough [22, Theorem 10.4], as explained below following an argument in [22, Section 10.4]:

Stability of the Periodic Orbit Through Averaging
Define the *averaged system*

$$\frac{d\bar{\Omega}}{d\theta}(\theta) = \epsilon\left(\frac{W_0}{\bar{\Omega}} - 1\right), \quad \text{where } W_0 = \frac{W_{\text{cycle}}}{2\pi k_B T_1} \tag{6.7}$$

is the dimensionless averaged work output over a complete rotation of the wheel. It follows that $\bar{\Omega} = W_0$ is an exponentially attractive equilibrium for the averaged system. Therefore, if $\bar{\Omega}(0) = W_0 \in [\Omega_{\text{min}}, \Omega_{\text{max}}]$ and $\Omega(0) - W_0 = O(\epsilon)$, there exists $\bar{\epsilon}$ such that for all $\epsilon \in (0, \bar{\epsilon})$, $\Omega(\theta)$ is defined and

$$\Omega(\theta) - W_0 = O(\epsilon) \qquad \text{for all } \theta \in [0, \infty).$$

Moreover, $\Omega(\theta)$ is a unique, exponentially stable, 2π-periodic solution of (6.6).

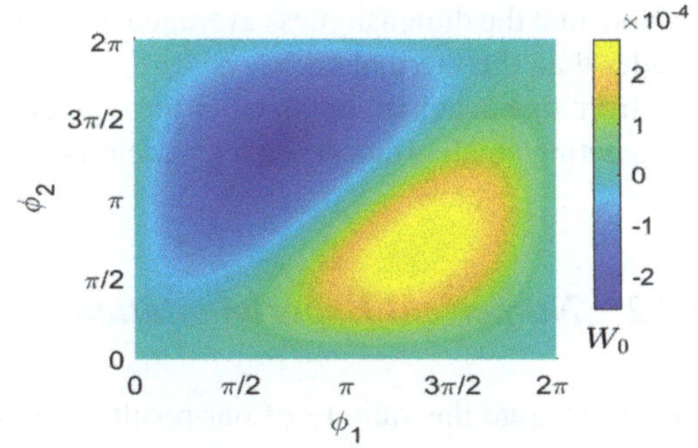

Fig. 6.4 Dimensionless quasi-static work over a cycle, W_0, for different values of the phase differences ϕ_1, ϕ_2. Note that the maximum of W_0 is found at the point $(\phi_1^*, \phi_2^*) \approx (4.25, 2.15)$. Image from [18]

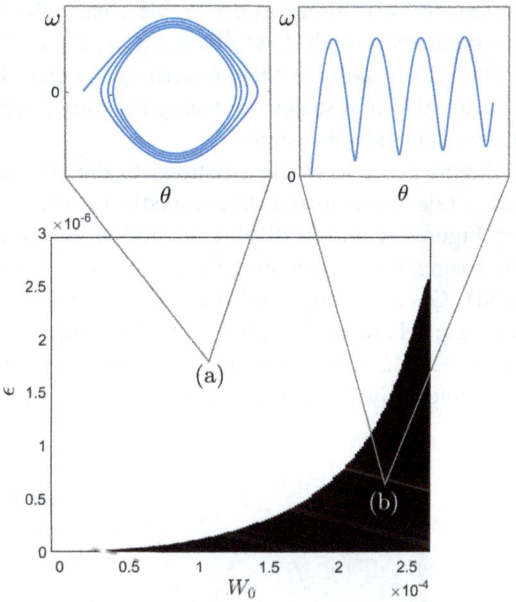

Fig. 6.5 Space of parameters (ϵ, W_0) numerically divided into two regions (by testing whether $\omega(\theta_0+2\pi) \geq \omega(\theta_0)$): (**a**) parameters for which the only stable solution is the stationary solution, and (**b**) both the stationary solution and the periodic solution are stable. In the blow-up figures we have, for each of the regions, an example trajectory in state-space. Image from [18]

Consequently, the wheel keeps rotating (i.e. $\Omega(\theta) > 0$) provided that W_0, the dimensionless averaged work output in (6.7), dominates the $O(\epsilon)$ discrepancy. Such a condition can be ensured by selecting the parameters ϕ_1 and ϕ_2 so that W_0 is positive, as can be seen in Fig. 6.4, and selecting ϵ sufficiently small. While a precise bound on the size of ϵ as a function of W_0 is not available, it can be numerically determined as shown in Fig. 6.5. The figure divides the space of parameters (ϵ, W_0) into two regions, depending on the type of equilibrium obtained. In region (a) the wheel will eventually come to a stall, regardless of the initial conditions, since dissipation (captured in ϵ) dominates the work production. On the other hand, for parameters in region (b), the wheel will settle to a periodic orbit, as long as the system starts with sufficient momentum, e.g., with initial conditions above the blue curve in the blow-up of Fig. 6.3.

Note that the dimensionless averaged work output W_0 is four orders of magnitude smaller than the thermal energy $\sim k_B T_1$. This, together with the continuous flux of heat from one heat bath to the other through the electrical degrees of freedom, leads to expecting small efficiencies, a characteristic typical of such engines.

6.2.2 Numerical Experimentation

To demonstrate the validity of our results, we numerically compute realizations of the original process (6.3) and compare them to the solutions of the deterministic system (6.5). In order to do so, we ensure the time-scale separation by selecting our parameters such that $\sqrt{R_1 R_2} C_0 \ll \Gamma/(k_B T_1)$ and $\sqrt{R_1 R_2} C_0 \ll I/\Gamma$, where $\sqrt{R_1 R_2} C_0$ is the time-scale governing the electrical subsystem, while $\Gamma/k_B T_1$ and I/Γ are the time-scales governing the oscillation and the damping, respectively, of the mechanical subsystem.

Figure 6.6 shows the solutions to the stochastic differential Eq. (6.3) in solid red, while those to the deterministic Eq. (6.5) are portrayed in dotted blue. The top Figures (a and b) display a solution for which the work output is positive and increasing, thus augmenting the wheel's velocity (at least until it reaches the periodic orbit). On the other hand, Figures (c) and (d) showcase the opposite situation, the case where the wheel eventually comes to a stall because the work output is not sufficient to overcome dissipation. To produce these plots we have used experimentally realizable parameters (given in Table 6.1), which are kept constant

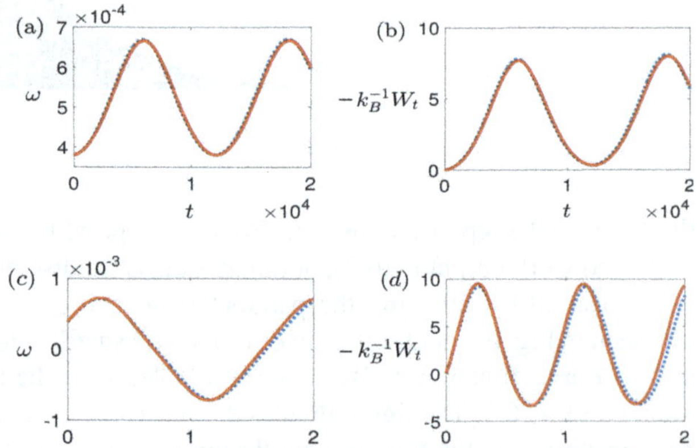

Fig. 6.6 Solutions to the stochastic system (6.3) are displayed in solid red, while those to the deterministic system (6.5) are shown in dotted blue. Figures (**a**) and (**b**) are the result of a successful choice of parameters in that work output is produced and the wheel's motion is enhanced. On the other hand, in Figures (**c**) and (**d**) the work production is not sufficient to overcome dissipation and thus the wheel oscillates around the static equilibrium point. Image from [18]

Table 6.1 Parameters in numerical simulation

Parameter	Value	Units
I	5×10^2	kg nm^2
Γ	10^{-2}	kg nm^2/s
ω_0	3.8×10^{-4}	1/s
θ_0	$\pi/2$	rad
t_f	2×10^4	s
R_1, R_2	1	Ω
T_1	200	K
T_2	400	K
k_B	10^{-23}	J/K
C_0	2	mF
β	0.1	–
ϕ_1	(a)&(b): 4.25 / (c)&(d): π	rad
ϕ_2	(a)&(b): 2.15 / (c)&(d): π	rad

except for ϕ_1 and ϕ_2, that vary from Figures (a) and (b) to Figures (c) and (d), from taking the optimal values to taking suboptimal ones. We remark that although the stochastic solution in solid red constitutes only one realization, the time-scale separation ensures that all realizations behave approximately like the deterministic solution, and, therefore, typical solutions to the stochastic equation do not differ much from one realization to another.

It is in order to point out the resemblance of the dynamical behavior of our stochastic heat engine with that of a damped pendulum with constant torque [21]. Indeed, due to the similarity of the shape of $F(\theta)$ to a shifted sine, it is of no surprise that both systems share the same type of equilibria in the different regions of their parameter space. However, the damped pendulum [21] has a qualitatively different behavior when the constant torque applied is higher than the amplitude of undulations of the potential. In such a case, only the periodic solution is a stable solution. We have not been able to replicate such a behavior in our stochastic heat engine, which can be attributed to the fact that the amplitude of $F(\theta)$ is always greater than its vertical shift due to restrictions of the work production.

Besides the resemblance of the engine dynamics to those of a damped forced pendulum, there is also a qualitative resemblance to the dynamical behavior of the Stirling engine as explained in [23, 24]. Specifically, a Stirling engine can be modeled as a periodic nonlinear pendulum, and its equilibrium modes have been experimentally shown to be the same as those of our stochastic heat engine [25]. Thus, the analysis presented herein can be used to establish the existence of periodic orbits for the macroscopic Stirling engine, and to identify conditions for which periodic motion persists.

6.3 Inertialess Gyrating Engine

We have seen how variations in the torque are averaged out by the inertia of the system, and that a considerable amount of inertia is required to produce limit cycle oscillations (see Fig. 6.5, where ϵ is inversely proportional to the inertia). While torque generating mechanisms are also ubiquitous in the biological world, inertia is not a property that one naturally associates with such processes. In the present section, seeking ways to dispense with the need for inertial effects, we study the combined effect of coupled Brownian gyrator engines. We show that an appropriate coupling averages out variations in the produced torque and helps overcome dissipative forces to allow sustained operation for vanishingly small inertia [6].

In fact, one can do so in greater generality by considering what we call gyrating engines, i.e., systems with a rotational degree of freedom characterized by an angle θ and driven by an external torque $\mathcal{T}$ that depends on θ, which, however, may not necessarily retain the same sign during a cycle. Specifically, the equations of motion for a gyrating engine are

$$\begin{aligned} \dot{\theta} &= \omega \\ I\dot{\omega} &= \mathcal{T}(\theta) - \Gamma\omega, \end{aligned} \tag{6.8}$$

where I is the moment of inertia and Γ is the friction coefficient. The term $-\Gamma\omega$ corresponds to external dissipation (or some torque proportional to angular velocity ω exchanged with an external subsystem acting as a load, as in the previous section). This model captures the general principle behind a wide range of mechanisms that convert thermal/chemical energy to rotary motion, whether synthetic or natural, from steam engines to biomolecular motors. Indeed, the Brownian gyrating engine presented in the previous chapter precisely obeys (6.8) with

$$\mathcal{T}^{\mathrm{B}}(\theta) = -\frac{1}{2}\,\mathrm{Tr}\left[\partial_\theta C^{-1}(\theta)\Sigma(\theta)\right].$$

The salient feature of this kind of devices is the inertia needed to average out fluctuations and ensure sustained operation.

Analogous biomolecular mechanisms, however, seem to dispense of such a need for inertial effects [26–28]. A cursory view of the workings of biomolecular engines reveals a many-fold symmetry of multiple torque-generating units at work. Specifically, there are three motor proteins that have been unambiguously identified as rotary engines, the $F0/F1$ ATP synthase and the bacterial flagellar motor [29]; they are powered by chemical gradients with the flagellar and $F0$ motor tapping onto trans-membrane ion-motive force while the $F1$ motor relying on ATP hydrolysis. Yet, in spite of great strides over the past 40 years into the workings of these $50nm$-scale motors, much remains to be understood [30]. In regard to the mechanics, their geometry, that engages several torque-generating subunits [31, 32] (up to

11 in flagellar motors, and often a threefold symmetry in ATPases), leads to the conclusion that a principle such as the one we present herein must be at work.

With this in mind, we study the coupling of multiple gyrating engines as a way to eliminate the need for inertia in sustained limit cycle oscillation. The basic idea explored herein is based on the principle that a *phase difference between coupled gyrating engines can average out the applied torque*. Thereby, angular variations in torque and load can be matched via a suitable geometric arrangement. Our interest is mainly in enabling sustained operation in the presence of sign-indefinite generated torque by individual engines, that is, in ensuring that the combined torque of multiple units retains its sign, which we prove ensures continuous operation.

6.3.1 Inertialess Forced Pendulum

Let us start by considering the damped pendulum with constant torque [33], i.e., a system that behaves according to (6.8) with

$$\mathcal{T}^{\mathrm{p}}(\theta) = \tau - \sin(\theta),$$

where τ represents the constant torque being applied. It is insightful to consider the effective potential that drives the motion,

$$U(\theta) = -\int_0^\theta \mathcal{T}(\vartheta)d\vartheta.$$

This has the form of a tilted sinusoid. A cartoon in two parts, corresponding to two different sets of parameters (of inertia and frictional forces), is displayed at the top row of Fig. 6.7. In this, the position of a ball rolling down the corrugated hill-side embodies the state of the pendulum; the drawing on the left exemplifies insufficient-inertia/excessive-friction for a limit cycle to exist, while the one on the right exemplifies a continuous operation. The second row of Fig. 6.7 depicts the collaborative effect of two coupled engines. In the analogy of two coupled balls, the combined center of gravity moves along a tilted straight line (red dashed line in the figure), facilitating the downward sliding along the periodic potential which is tilted due to the applied torque.

The situation with Brownian gyrating engine is analogous. The coupling of a number of engines, with a suitable phase difference between one another, averages out the "bumps" in the "corrugated" potential and enables sustained operation for a vanishingly small applied torque. To make this explicit, let us begin by highlighting the effect of coupling several damped pendula with an applied constant torque and a certain phase difference. Specifically, for this case, we consider two pendula coupled with a phase difference of π radians (see Fig. 6.7, bottom). The effective torque on the combined system is

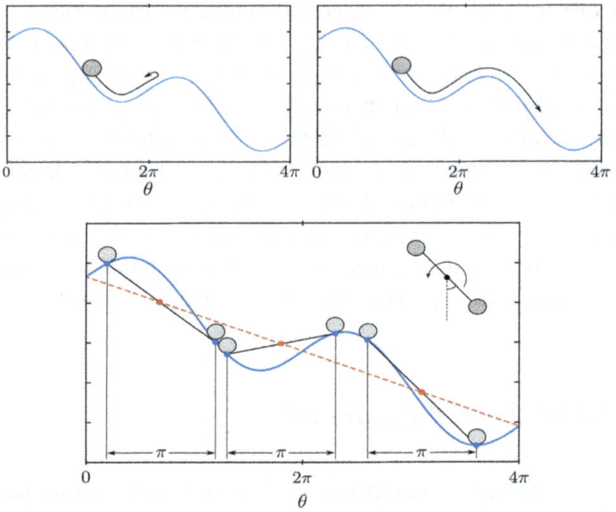

Fig. 6.7 Potential for the damped pendulum with constant torque. Two cases are displayed. Top-left: inertial effects are not able to overcome the uphills generated by gravity and the only stable solution is the stationary one. Top-right: both inertial effects and constant torque (slope) are enough to sustain continuous motion and the pendulum reaches a stable periodic orbit. Bottom: the average of two potentials displaced by a π phase difference is linear in θ. The graphic representation provides insight into how two θ-equispaced coupled pendula with a constant torque operate stably in a limit cycle: their combined effective potential is a sloped line (red dashed line in the figure). Image from [6]

$$\mathcal{T}_2^{\mathrm{p}}(\theta) = \frac{1}{2}\left(\mathcal{T}^{\mathrm{p}}(\theta) + \mathcal{T}^{\mathrm{p}}(\theta+\pi)\right) = \tau - \frac{1}{2}\left(\sin(\theta) + \sin(\theta+\pi)\right) = \tau,$$

effectively canceling the undulations of the potential; the $\frac{1}{2}$ factor scales the power of the two engines so that they can be compared to one engine. Thus, the effective torque remains constant, and thereby the overall potential driving the system of two engines has a constant tilt with no undulations. The system requires neither any inertia nor a minimum amount of actuation to achieve sustained continuous rotation. The cartoon shown in Fig. 6.7 helps exemplify the effect.

The underlying principle is readily seen to rely on ensuring the sign-definiteness of the effective torque. This is carried out via cancellation of respective terms between the Fourier series expansion of applied torques from contributing units. The sign-definiteness of the effective torque guarantees stable limit cycle oscillation, as is shown in Appendix C.1. Evidently, in more complicated examples, higher-order harmonics are not immune and can likewise be eliminated or suppressed by coupling more engines as shown in the analysis that follows.

6.3.2 Inertialess Brownian Gyrating Engine

We consider the equidistant (in the θ space) coupling of two and three Brownian gyrating engines, that generate combined torque

$$\mathcal{T}_2^{\mathrm{B}}(\theta) = \frac{1}{2}\left(\mathcal{T}^{\mathrm{B}}(\theta) + \mathcal{T}^{\mathrm{B}}(\theta + \pi)\right), \text{ and}$$

$$\mathcal{T}_3^{\mathrm{B}}(\theta) = \frac{1}{3}\left(\mathcal{T}^{\mathrm{B}}(\theta) + \mathcal{T}^{\mathrm{B}}(\theta + 2\pi/3) + \mathcal{T}^{\mathrm{B}}(\theta + 4\pi/3)\right),$$

for the two- and three-engine configuration, respectively. The resulting potential U is shown in the insert in Fig. 6.8 over two periods. It changes from a periodic slopped shape (in the case of one engine), to practically a slopped straight line already for two coupled engines, and more so for three. The main plot in Fig. 6.8 shows the averaged steady state angular velocity as a function of inertia. It is seen that, for this set of parameters (which are given in Table 6.2), three engines dispense completely of the need for inertia, ensuring a limit cycle; with two engines the need for inertia is already minimal.

Figure 6.9 illustrates how the averaged final angular velocity varies with the temperature difference $\Delta T := T_2 - T_1$ that powers the gyrator. Evidently, the coupling of multiple Brownian gyrating engines reduces the threshold temperature

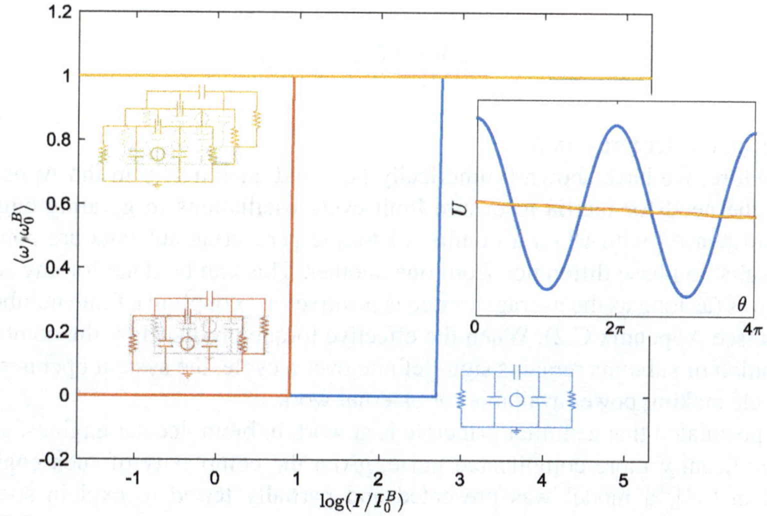

Fig. 6.8 Left: normalized averaged final angular velocity $\langle\omega/\omega_0^B\rangle$ vs $\log(I/I_0^B)$, with $\Delta T = 10$K for one, two and three coupled Brownian gyrating engines, respectively. Here, $I_0^B = \Gamma/\omega_0^B$ and ω_0^B is as defined in (6.9). Right: effective potential along two cycles. Image from [6]

Table 6.2 Parameters used in numerical simulations

Brownian gyrator engine		
Parameter	Value	Units
C_0	2	mF
β	0.1	–
R_1, R_2	1	Ω
T_1	200	K
k_B	$1.38 \cdot 10^{-23}$	$\mathrm{kg\,m^2\,s^{-2}\,K^{-1}}$
I	10^{-12} to 10^{-19} (Fig. 6.8)	$\mathrm{kg\,m^2}$
	$5 \cdot 10^{-16}$ (Fig. 6.9)	
ΔT	10 (Fig. 6.8)	K
	0–15 (Fig. 6.9)	
Γ	$4.32 \cdot 10^{-22}$	$\mathrm{kg\,m^2\,s^{-1}}$

difference needed for continuous operation. When three engines are coupled, the threshold is virtually eliminated, guaranteeing the existence of a limit cycle for vanishingly small ΔT. In Appendix C.2.1 we show that a sufficient condition for the torque $\mathcal{T}_3^B(\theta)$ to be always positive is

$$\frac{\sqrt{3}}{64}\frac{\Delta T}{\bar{T}} \gg \beta$$

where $\bar{T} = (T_1 + T_2)/2$. Then, we can approximate the average angular velocity as

$$\langle \omega \rangle \approx \frac{\sqrt{3} k_B \Delta T}{64\Gamma} \beta^2 =: \omega_0^B, \tag{6.9}$$

up to second-order terms in β.

Therefore, we have shown (numerically here and analytically in the Appendix C) that the need for inertia to ensure limit cycle oscillations in gyrating engines can be dispensed with when a number of torque-generating subunits are coupled with a suitable phase difference from one another. This can be done for any set of parameters (as long as the average torque is positive), by coupling a finite number of engines (see Appendix C.2). When the effective torque produced by the combined contribution of subunits remains sign-definite over a cycle, the system operates in a limit cycle making power available for external work.

It is postulated that a similar principle is at work in biomolecular engines, albeit in a significantly more complicated guise, given the complexity of such engines. Indeed, in [34], a model was presented and partially tested to explain specific physical mechanisms for torque generation in bacterial flagellar motors (BFMs). In this, a number of torque generating units with a "wide and gently slopping energy well" contribute in ways that are reminiscent of the principle presented herein. Although the physics of torque generation remain poorly understood, it was proposed in [34] that both electrostatic and steric forces are at work, with the latter

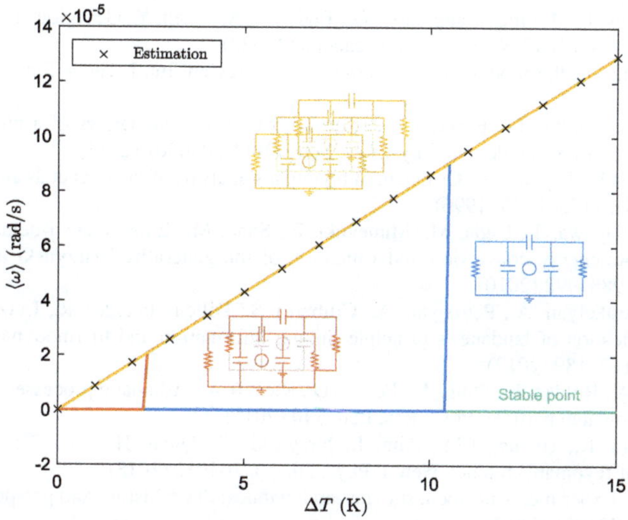

Fig. 6.9 Average limit cycle angular velocity $\langle\omega\rangle$ vs. ΔT for one, two and three coupled Brownian gyrating engines. An estimation of the average angular velocity from (6.9) has been marked by black "×", matching the numerical results. The (flat) green line corresponds to a stable equilibrium present when the effective torque fails to be sign-definite. Image from [6]

generating a "push". The resulting torque profile may likely necessitate multiple units to smooth out higher harmonics that may thus be present. Understanding how ion-driven molecular machines work is of fundamental importance in cellular biology, and thus it is likely that the principle discussed herein may help explain the workings of multiple torque-generating subunits [31, 32] and, perhaps, even the necessity for a large number (up to 11 in flagellar motors) of such units for the corresponding torque-generating potential.

References

1. Battle, C., Broedersz, C.P., Fakhri, N., Geyer, V.F., Howard, J., Schmidt, C.F., MacKintosh, F.C.: Broken detailed balance at mesoscopic scales in active biological systems. Science (2016)
2. Gnesotto, F.S., Mura, F., Gladrow, J., Broedersz, C.P.: Broken detailed balance and non-equilibrium dynamics in living systems: a review. Reports on Progress in Physics. IOP publishing (2018)
3. Chiang, K.-H., Lee, C.-L., Lai, P.-Y., Chen, Y.-F.: Electrical autonomous Brownian gyrator. Phys. Rev. E **96**, 032123 (2017)
4. Ciliberto, S., Imparato, A., Naert, A., Tanase, M.: Heat flux and entropy produced by thermal fluctuations. Phys. Rev. Lett. **110**, 180601 (2013)
5. Ciliberto, S., Imparato, A., Naert, A., Tanase, M.: Statistical properties of the energy exchanged between two heat baths coupled by thermal fluctuations. J. Stat. Mech. Theory Exp. **2013**(12), P12014 (2013)

6. Ventura Siches, J., Movilla Miangolarra, O., Taghvaei, A., Chen, Y., Georgiou, T.T.: Inertialess gyrating engines. PNAS Nexus. Oxford academic (2022)
7. Bennett, C.H.: The thermodynamics of computation–a review. Int. J. Theor. Phys. **21**(12), 905–940 (1982)
8. Strasberg, P., Schaller, G., Brandes, T., Esposito, M.: Thermodynamics of a physical model implementing a maxwell demon. Phys. Rev. Lett. **110**(4), 040601 (2013)
9. Parrondo, J.M.R., Español, P.: Criticism of feynman's analysis of the ratchet as an engine. Am. J. Phys. **64**(9), 1125–1130 (1996)
10. Toyabe, S., Sagawa, T., Ueda, M., Muneyuki, E., Sano, M.: Experimental demonstration of information-to-energy conversion and validation of the generalized jarzynski equality. Nat. Phys. **6**(12), 988–992 (2010)
11. Bérut, A., Arakelyan, A., Petrosyan, A., Ciliberto, S., Dillenschneider, R., Lutz, E.: Experimental verification of landauer's principle linking information and thermodynamics. Nature **483**(7388), 187–189 (2012)
12. Martínez, I.A., Roldán, E., Dinis, L., Petrov, D., Rica, R.A.: Adiabatic processes realized with a trapped Brownian particle. Phys. Rev. Lett. **114** (2015)
13. Bang, J., Pan, R., Hoang, T.M., Ahn, J., Jarzynski, C., Quan, H.T., Li, T.: Experimental realization of feynman's ratchet. New J. Phys. **20**(10), 103032 (2018)
14. Ciliberto, S.: Experiments in stochastic thermodynamics: Short history and perspectives. Phys. Rev. X **7**(2), 021051 (2017)
15. Lin, W., Liao, Y.-H., Lai, P.-Y., Jun, Y.: Stochastic currents and efficiency in an autonomous heat engine. Phys. Rev. E **106**, L022106 (2022)
16. Serra-Garcia, M., Foehr, A., Molerón, M., Lydon, J., Chong, C., Daraio, C.: Mechanical autonomous stochastic heat engine. Phys. Rev. Lett. **117**, 010602 (2016)
17. Bryant, S.J., Machta, B.B.: Energy dissipation bounds for autonomous thermodynamic cycles. Proc. Natl. Acad. Sci. **117**(7), 3478–3483 (2020)
18. Movilla Miangolarra, O., Taghvaei, A., Chen, Y., Georgiou, T.T.: Thermodynamic engine powered by anisotropic fluctuations. Phys. Rev. Res. **4**, 023218 (2022)
19. Movilla Miangolarra, O., Taghvaei, A., Fu, R., Chen, Y., Georgiou, T.T.: Energy harvesting from anisotropic fluctuations. Phys. Rev. E (2021)
20. Movilla Miangolarra, O., Taghvaei, A., Chen, Y., Georgiou, T.T.: Geometry of finite-time thermodynamic cycles with anisotropic thermal fluctuations. IEEE Control Syst. Lett. **6**, 3409–3414 (2022)
21. Pierre, C., Gilli, J.M., Monticelli, M., Vanderberghe, N.: A damped pendulum forced with a constant torque. Am. J. Phys. **73** (2005)
22. Khalil, H.K.: Nonlinear Control. Pearson Higher Education, Upper Saddle River (2014)
23. Izumida, Y.: Nonlinear dynamics analysis of a low-temperature-differential kinematic stirling heat engine. Europhys. Lett. **121**(5), 50004 (2018)
24. Izumida, Y.: Quasilinear irreversible thermodynamics of a low-temperature-differential kinematic stirling heat engine. Phys. Rev. E **102**, 012142 (2020)
25. Toyabe, S., Izumida, Y.: Experimental characterization of autonomous heat engine based on minimal dynamical-system model. Phys. Rev. Res. **2**, 033146 (2020)
26. Meister, M., Lowe, G., Berg, H.C.: The proton flux through the bacterial flagellar motor. Cell **49**(5), 643–650 (1987)
27. Noji, H., Yasuda, R., Yoshida, M., Kinosita, K.: Direct observation of the rotation of F1-ATPase. Nature **386**(6622), 299–302 (1997)
28. Kay, E.R., Leigh, D.A., Zerbetto, F.: Synthetic molecular motors and mechanical machines. Angew. Chem. Int. Ed. **46**(1–2), 72–191 (2007)
29. Oster, G., Wang, H.: Rotary protein motors. Trends Cell Biol. **13**(3), 114–121 (2003)
30. Sowa, Y., Berry, R.M.: Bacterial flagellar motor. Q. Rev. Biophys. **41**(2), 103–132 (2008)
31. Reid, S.W., Leake, M.C., Chandler, J.H., Lo, C.-J., Armitage, J.P., Berry, R.M.: The maximum number of torque-generating units in the flagellar motor of escherichia coli is at least 11. Proc. Natl. Acad. Sci. **103**(21), 8066–8071 (2006)

32. Jia, Y., Li, J.: Reconstitution of F0F1-ATPase-based biomimetic systems. Nat. Rev. Chem. **3**(6), 361–374 (2019)
33. Coullet, P., Gilli, J.M., Monticelli, M., Vandenberghe, N.: A damped pendulum forced with a constant torque. Am. J. Phys. **73**(12), 1122–1128 (2005)
34. Mandadapu, K.K., Nirody, J.A., Berry, R.M., Oster, G.: Mechanics of torque generation in the bacterial flagellar motor. Proc. Natl. Acad. Sci. **112**(32), E4381–E4389 (2015)

Conclusions

In these pages, we have attempted to grasp the intrinsic idiosyncrasies of thermodynamic systems subject to thermal anisotropy. Within this context, we have touched upon the core problems in Stochastic Thermodynamics such as minimizing entropy production and quantifying the arrow of time, or optimally extracting energy both externally and autonomously, while exploiting a geometric view that hinges upon Optimal Mass Transport. In doing so, we have tried to draw, as much as possible, a parallel between the standard problems in Stochastic Thermodynamics presented in the introduction (Chap. 2), and their anisotropic counterpart (in particular Chaps. 3, 4 and 5).

It is now in order to highlight their differences in a way, we hope, clarifies the nature of thermal anisotropy. First, in Chap. 4, perhaps the most obvious characteristic of anisotropic systems becomes apparent: the anisotropy in fluctuations can be exploited to extract work from the system, without requiring switching between heat baths or changing their temperature as in the isotropic case. Therein, a geometric approach to optimal work extraction, that results in an isoperimetric problem in the space of thermodynamic states, is provided, endowing the problem with a rich structure analogous to that of classical thermodynamics where work can be expressed as an area integral over the parameter space. This is materialized in the autonomous actuation of a microscopic engine in Chap. 6, which continuously harvests energy in a way reminiscent of certain biomolecular engines.

In Chap. 5, we show that access to divergence-free (non-conservative) actuation is vital when attempting to stall the entropy production associated with thermal anisotropy. This can be traced to the fact that optimal velocities that solve the weighted Optimal Mass Transport problem are no longer of gradient form, but of $T\nabla\phi$ form. This is in striking contrast to the standard (isotropic) case in which optimal velocities can be obtained by control on the potential, highlighting the distinctive unavoidability of entropy production for systems in simultaneous contact with multiple heat baths that relax to non-equilibrium steady-states. We have found a meaningful way to decompose entropy production in which the asymmetry induced

O. Movilla Miangolarra, *Stochastic Thermodynamic Treatment of Thermal Anisotropy*, Springer Theses, https://doi.org/10.1007/978-3-031-68066-3

by temperature plays a central role. This decomposition separates the entropy production we are familiar with (excess), from the one induced by anisotropy (housekeeping), leading to several optimization approaches for minimizing entropy production that trade off housekeeping and excess entropy production. Rather surprisingly, vanishing housekeeping entropy producing non-equilibrium trajectories are shown to exist even if we only have control over conservative forcing.

Another central feature of anisotropy is that entropy and energy require different treatments since they typically differ by a factor of T, which is now a matrix. This results in the need for different metrics when measuring energy dissipated and entropy production (compare (4.31) to (5.6)), and it is precisely this mismatch what allows for energy extraction. Indeed, this difference between metrics is highlighted in the second law, where optimal protocols (those that saturate the inequality) differ whether we write the second law in terms of entropy (5.6) or of (free) energy (4.31). Thus, one needs to choose the most meaningful metric for each particular problem, in contrast to the isotropic case where both metrics and both versions of the second law coincide. Moreover, matricial temperatures lead to ways of decomposing entropy, intrinsically different from those of decomposing energy; contrary to the standard case in which the non-equilibrium steady-state is induced by non-conservative forces and a similar decomposition can be used for both [1, 2]. The discrepancy between energy and entropy becomes apparent in the definition of free energy, typically given by $F = E - k_B T S$. The temperature being now a matrix makes evident the need for a change of definition and the general difficulty of switching between energies and entropies.

This brings our study of thermodynamic systems in thermally anisotropic environments to its end. Yet, a lot remains to be understood. First, the assumption of a two-dimensional parametrization of the thermodynamic state has been necessary for results such as the isoperimetric problem of energy harvesting or the usage of the isoperimetric inequality to bound efficiency. The reason for this assumption is the lack of simple geometric tools in higher/infinite-dimensional manifolds of thermodynamic states, such as a pertinent measure of area. This is the subject of ongoing work.

Second, we have considered the problem of maximizing work output over a cycle through conservative actuation only. Relaxing this condition to also account for non-conservative forcing (outside the steady-state regime) would unveil an interesting trade-off between the isoperimetric cycles that maximize work extraction through conservative forcing, and cycles resulting in a circulation in the velocity field that maximizes power extraction through non-conservative forcing. On the other hand, we have also restricted ourselves to the overdamped Langevin equation model. There is no doubt in mind that understanding anisotropy in other models would shed more light on the problem. Especially captivating seems the search for a geometric view of entropy production in underdamped thermodynamic systems [3, 4].

From the control theoretic point of view, conditions pertaining to the existence of optimal controls and robustness to uncertainty in the constituents and environment have been left aside, not for lack of importance but for limited time and scope. Similarly, although the stability of work extracting limit cycles has been addressed when possible, we would like to emphasize that more attention should be paid to this subject. Indeed, the physics community has typically overlooked the stability of limit cycles, even if experimentally feasible cycles always require this property.

Ultimately, we hope this endeavor will serve as a stepping stone in understanding the ways and characteristics of anisotropy, whether it be of chemical potentials, particle concentration or temperature fields. We believe this is of prime importance, as the universal source of energy it is, in understanding how living systems are fueled, how we evolved to exploit these resources and how life manages to sustain itself against all odds. It is anisotropy that has allowed us to deceive the second law long enough to create a rich and vital oasis, that survives despite the clenching jaw of entropy production and the sharp claws of our unstoppable future. A future that will inevitably lead us to homogeneous, isotropic temperatures and lifeless lumps of matter.

References

1. Dechant, A., Sasa, S.-I., Ito, S.: Geometric decomposition of entropy production in out-of-equilibrium systems. Phys. Rev. Res. (2022)
2. Hatano, T., Sasa, S.: Steady-state thermodynamics of Langevin systems. Phys. Rev. Lett. **86** (2001)
3. Dechant, A., Sakurai, Y.: Thermodynamic interpretation of Wasserstein distance. Preprint (2019). arXiv:1912.08405
4. Sabbagh, R., Movilla Miangolarra, O., Georgiou, T.T.: Wasserstein speed limits for Langevin systems. Phys. Rev. Res. **6**(3), 033308 (2024)

Appendix A
Underdamped Stochastic Heat Engine Subject to a Time-Varying Temperature

Energy and entropy exchange between a heat bath and a particle ensemble is not particular to Carnot engines (like the one presented in Sect. 2.4). In fact, energy exchanges are omnipresent in biological engines, where they appear to be mediated by continuous processes and energy differentials [1], rather than abrupt changes such as those associated with the classical Carnot cycle.

In this appendix, we consider the rudimentary model of a cyclic process (periodic temperature profile of a heat bath) that fuels a colloidal engine following the Langevin dynamics [2]. The periodic temperature profile may stem from a non-equilibrium (or active) heat bath; a heat bath constituted by self-propelled particles (e.g. bacteria) with periodically varying activities [3], or may be externally engineered [4]. Our goal is to optimize the performance of such an engine. Specifically, we derive explicit bounds as well as optimal control protocols that draw maximum power and achieve maximum efficiency at any specified level of power. Hence, the work presented below is an extension of recent studies for periodic temperature profile in the linear regime [5–7]. It addresses the nonlinear regime and, in particular, the case of underdamped dynamics with low friction.

A.1 Energetics and Low Friction Approximation

Throughout this appendix, we will consider the underdamped Langevin dynamics (2.1) with a harmonic potential

$$U(t, x) = \frac{1}{2} q(t) x^2,$$

where $q(t)$ represents its intensity and constitutes our control variable. Since a harmonic potential leads to linear Langevin dynamics (2.1), it is sufficient to

O. Movilla Miangolarra, *Stochastic Thermodynamic Treatment of Thermal Anisotropy*, Springer Theses, https://doi.org/10.1007/978-3-031-68066-3

consider the following governing equations for the variance (see Sect. 3.4 for details on the derivation):

$$\dot{\Sigma}_x(t) = 2\Sigma_{xv}(t), \tag{A.1a}$$

$$\dot{\Sigma}_{xv}(t) = \Sigma_v(t) - \frac{q(t)}{m}\Sigma_x(t) - \frac{\gamma}{m}\Sigma_{xv}(t), \tag{A.1b}$$

$$\dot{\Sigma}_v(t) = -\frac{2q(t)}{m}\Sigma_{xv}(t) - \frac{2\gamma}{m}\Sigma_v(t) + \frac{2\gamma k_B}{m^2}T(t), \tag{A.1c}$$

where Σ_x, Σ_v are the variances of X_t and v_t, respectively, while $\Sigma_{xv}(t)$ is the correlation between X_t and v_t.

The total energy of such an ensemble can be written as the sum of potential and kinetic energies, namely,

$$\mathcal{E} = \mathbb{E}\left[\frac{1}{2}mv_t^2 + \frac{1}{2}q(t)X_t^2\right] = \frac{1}{2}m\Sigma_v + \frac{1}{2}q(t)\Sigma_x.$$

Its differential is given by

$$d\mathcal{E} = \frac{1}{2}\dot{q}(t)\Sigma_x dt - \gamma\Sigma_v dt + \frac{\gamma k_B T(t)}{m}dt,$$

and can be split into two contributions: one stemming from the variation of our control parameter, representing work

$$đ\mathcal{W} = \frac{1}{2}\dot{q}(t)\Sigma_x(t)dt,$$

and another due to the interaction between the system and the environment,

$$đ\mathcal{Q} = -\gamma\Sigma_v(t)dt + \frac{\gamma k_B T(t)}{m}dt, \tag{A.2}$$

which represents heat.

Although the differential equations for the evolution of the variance are explicit and in closed form, the analysis for maximizing power and efficiency is still challenging due to the presence of the three coupled ODEs. As a result, it is common to consider simplifying approximations that hold in different physical regimes. Our analysis will concern the low friction regime studied in [8, 9]. When the temperature $T(t)$ and the harmonic potential intensity $q(t)$ are constant, the low friction regime holds if the particle in (2.1) undergoes many oscillations before dissipation effects become significant. Mathematically, this amounts to $\sqrt{\frac{q}{m}} \gg \frac{\gamma}{m}$. In the case where $T(t)$ and $q(t)$ are time varying periodic functions, it is required in addition that their power spectrum mostly lies below the natural frequency of oscillations $\sqrt{q/m}$.

Under the low friction assumptions, the equipartition condition

$$\frac{1}{2}q(t)\Sigma_x(t) = \frac{1}{2}m\Sigma_v(t) \tag{A.3}$$

holds approximately [9, Section 4.2], which implies energy is equally distributed between the position and velocity degrees of freedom. As a result, it is possible to express the dynamics of our system as a function of $\Sigma_v(t)$ uniquely. More precisely, we can use (A.1a) to express $\Sigma_{xv}(t)$ as follows:

$$\Sigma_{xv}(t) = \frac{1}{2}\dot{\Sigma}_x = \frac{m}{2q(t)}\dot{\Sigma}_v - \frac{m\dot{q}(t)}{2q(t)^2}\Sigma_v,$$

where the second equality comes from the equipartition condition (A.3). Now, we can rewrite $\Sigma_{xv}(t)$ in (A.1c), to obtain

$$\dot{\Sigma}_v(t) = \Sigma_v(t)\left(\frac{\dot{q}(t)}{2q(t)} - \frac{\gamma}{m}\right) + \frac{\gamma k_B}{m^2}T(t). \tag{A.4}$$

For this equation to be meaningful, we require $q(t)$ to be (weakly) differentiable. Equation (A.4) is the underlying model for our analysis through the rest of this appendix.

A.2 Analysis of Maximum Power

We consider a stochastic thermodynamic engine driven by a periodic temperature $T(t)$ with period t_f. The averaged power output over a cycle is equal to

$$\mathcal{P} = -\frac{1}{t_f}\int_0^{t_f} đ\mathcal{W}.$$

Using the first law of thermodynamics $d\mathcal{E} = đ\mathcal{Q} + đ\mathcal{W}$, the power can be expressed in terms of heat exchange as follows,

$$\mathcal{P} = \frac{1}{t_f}\int_0^{t_f}(đ\mathcal{Q} - d\mathcal{E}) = \frac{\gamma}{t_f}\int_0^{t_f}\left(\frac{k_B T(t)}{m} - \Sigma_v(t)\right)dt + \frac{1}{t_f}(\mathcal{E}(t_f) - \mathcal{E}(0)), \tag{A.5}$$

where the definition (A.2) is used and $\mathcal{E}|_t \equiv \mathcal{E}(t)$, for clarity. To account for possible discontinuities in $T(t)$ and $q(t)$, as it is often the case in the underdamped setting [8, 10, 11], we consider a cycle from 0^+ to t_f^+. Thus, q, Σ_v and $\mathcal{E}$, which are periodic, satisfy

$$q(0^+) = q(t_f^+) \quad \text{and} \quad \Sigma_\nu(0^+) = \Sigma_\nu(t_f^+), \tag{A.6}$$

and $\mathcal{E}(0^+) = \mathcal{E}(t_f^+)$. The latter condition removes the boundary terms in (A.5), while conditions (A.6) impose an integral constraint on $\Sigma_\nu(t)$. Specifically, if we divide (A.4) by $\Sigma_\nu(t)$ and integrate from 0^+ to t_f^+ we obtain

$$\log\left(\frac{\Sigma_\nu(t_f^+)}{\Sigma_\nu(0^+)}\right) = \log\left(\frac{q(t_f^+)}{q(0^+)}\right)^{\frac{1}{2}} - \frac{\gamma}{m}t_f + \frac{\gamma k_B}{m^2}\int_0^{t_f}\frac{T(t)}{\Sigma_\nu(t)}dt. \tag{A.7}$$

Thus, in view of (A.6),

$$\int_0^{t_f}\frac{T(t)}{\Sigma_\nu(t)}dt = \frac{mt_f}{k_B}. \tag{A.8}$$

This integral constraint is the main ingredient in our analysis for both maximum power and maximum efficiency.

We first consider maximizing power output, namely,

$$\max_{\Sigma_\nu}\frac{\gamma}{t_f}\int_0^{t_f}\left(\frac{k_B T(t)}{m} - \Sigma_\nu(t)\right)dt, \tag{A.9}$$

subject to the constraint (A.8). We observe that due to (A.8),

$$\int_0^{t_f}\Sigma_\nu(t)dt = \int_0^{t_f}\left(\Sigma_\nu(t) + c\frac{T(t)}{\Sigma_\nu(t)}\right)dt - c\frac{mt_f}{k_B}$$

for any arbitrary constant c. The minimal value for the integrand $\Sigma_\nu(t) + c\frac{T(t)}{\Sigma_\nu(t)}$, pointwise and with respect to $\Sigma_\nu(t)$, is attained for $\Sigma_\nu(t) = \sqrt{cT(t)}$. This choice for $\Sigma_\nu(t)$ also satisfies (A.8) for $\sqrt{c} = \frac{k_B}{mt_f}\int_0^{t_f}\sqrt{T(t)}dt$. Therefore,

$$\Sigma_\nu(t) = \left(\frac{k_B}{mt_f}\int_0^{t_f}\sqrt{T(s)}ds\right)\sqrt{T(t)} \tag{A.10}$$

is the unique maximizer of (A.9).

The relation between $\Sigma_\nu(t)$ and the protocol $q(t)$ can be obtained from (A.4) by direct integration (cf. (A.7)), which for the optimizing $\Sigma_\nu(\cdot)$ gives

$$q(t) = q_0\frac{T(t)}{T(0)}\exp\left(\frac{2\gamma t_f}{m}\left(\frac{t}{t_f} - \frac{\int_0^t\sqrt{T(s)}ds}{\int_0^{t_f}\sqrt{T(s)}ds}\right)\right). \tag{A.11}$$

Note that, q_0 specifies the starting value $q(0)$ of the optimal protocol. Further, $q(t)$ is continuous at all times when $T(t)$ is, which agrees with the intuition that jumps in the control are necessitated for adapting to a sudden jump in temperature [8, 12].

Finally, inserting (A.10) into the expression for power, we obtain that, under this protocol, any value for q_0 gives the same level of power, namely,

$$\mathcal{P}^* = \frac{\gamma k_B}{m}\mathrm{Var}(\sqrt{T}), \tag{A.12}$$

where

$$\mathrm{Var}(\sqrt{T}) = \frac{1}{t_f}\int_0^{t_f} T(t)dt - \left(\frac{1}{t_f}\int_0^{t_f}\sqrt{T(t)}dt\right)^2$$

quantifies fluctuations in $\sqrt{T(t)}$. Thus, expression (A.12) quantifies the maximum power that can be drawn in terms of an explicit measure of temperature fluctuations. Let us recap these results in the following statement:

Maximum Power for an Underdamped System Under an Arbitrary Temperature Profile

Consider an underdamped particle under a quadratic potential for which the velocity covariance follows (A.4). Then, after a cycle over $[0, t_f]$, the maximum power is

$$\mathcal{P}^* = \frac{\gamma k_B}{m}\mathrm{Var}(\sqrt{T}),$$

where

$$\mathrm{Var}(\sqrt{T}) = \frac{1}{t_f}\int_0^{t_f} T(t)dt - \left(\frac{1}{t_f}\int_0^{t_f}\sqrt{T(t)}dt\right)^2.$$

This power can be achieved by the optimal protocol

$$q(t) = q_0\frac{T(t)}{T(0)}\exp\left(\frac{2\gamma t_f}{m}\left(\frac{t}{t_f} - \frac{\int_0^t\sqrt{T(s)}ds}{\int_0^{t_f}\sqrt{T(s)}ds}\right)\right).$$

It is interesting to consider the shape of the temperature profile that allows maximum power to be drawn under the optimal protocol. In view of the above, it is easy to see that when $T(t) \in [T_c, T_h]$, the Carnot temperature profile

$$T_{\mathrm{Carnot}}(t) = \begin{cases} T_h, & t \in [0, \frac{t_f}{2}) \\ T_c, & t \in [\frac{t_f}{2}, t_f), \end{cases} \tag{A.13}$$

allows drawing maximum power, which in this case is

$$\frac{\gamma k_B}{m}\frac{(\sqrt{T_h}-\sqrt{T_c})^2}{4}. \tag{A.14}$$

This special profile (Carnot) has already been studied in [8], where result (A.14) has been obtained.

A.3 Analysis of Maximum Efficiency at Fixed Power

Let us now consider the efficiency,

$$\eta = \frac{\mathcal{W}_{\text{out}}}{\mathcal{Q}_{\text{qs}}},$$

where $\mathcal{Q}_{\text{qs}} = \mathcal{W}_{\text{out}} + \mathcal{W}_{\text{diss}}$. The dissipated work can be derived by multiplying the total entropy production by the temperature as

$$T(t)\dot{\mathcal{S}}_{\text{tot}} = T(t)\dot{\mathcal{S}}_{\text{sys}} - \dot{\mathcal{Q}} = T(t)\dot{\mathcal{S}}_{\text{sys}} + \dot{\mathcal{W}} - \dot{\mathcal{E}} = \dot{\mathcal{W}}_{\text{diss}},$$

where for the second equality we have used the first law, while for the third equality we have used the definition of dissipated work (2.18). Then, integrating both sides of the last equality over the time interval, and noting that the total energy does not change over a cycle, we obtain

$$\mathcal{Q}_{\text{qs}} = -\mathcal{W} + \mathcal{W}_{\text{diss}} = \int_0^{t_f} T(t)\dot{\mathcal{S}}_{\text{sys}} dt,$$

(also, $\mathcal{Q}_{\text{qs}} = -\int_0^{t_f} \mathcal{S}_{\text{sys}}\dot{T}(t)dt$, using integration by parts). It represents the effective uptake of thermal energy, which can also be thought of as the maximal amount of work that can be extracted when the dissipation is zero and the transition takes place quasi-statically [7].

In the underdamped low friction regime with quadratic potential, the distribution ρ is Gaussian with independent components (x, v) whose covariance are $\Sigma_x(t)$ and $\Sigma_v(t)$, respectively. Using the standard formula for the entropy of Gaussian distributions together with the equipartition condition (A.3) expressed in terms of the variances, $\mathcal{Q}_{\text{qs}}$ simplifies to

$$\mathcal{Q}_{\text{qs}} = -\frac{k_B}{2}\int_0^{t_f} \ln\left((2\pi e)^2 \frac{m}{q(t)} \Sigma_v^2(t)\right)\dot{T}(t)dt.$$

In the case where $T(t)$ is discontinuous, the limits of integration need to be taken as 0^+ and t_f^+, and similarly in what follows. Integration by parts gives

$$\begin{aligned}\mathcal{Q}_{\rm qs} &= \frac{k_B}{2}\int_0^{t_f}\left(2\frac{\dot{\Sigma}_\nu(t)}{\Sigma_\nu(t)} - \frac{\dot{q}(t)}{q(t)}\right)T(t)dt \\ &= \frac{k_B^2\gamma}{m^2}\int_0^{t_f}\frac{T(t)^2}{\Sigma_\nu(t)}dt - \frac{k_B\gamma}{m}\int_0^{t_f}T(t)dt,\end{aligned}$$

where we have used (A.4) to establish the second equality.

We now seek to optimize efficiency at a given power. To this end, it suffices to minimize the effective heat uptake $\mathcal{Q}_{\rm qs}$ subject to the given power. Thus, we consider

$$\min_{\Sigma_\nu}\int_0^{t_f}\frac{T(t)^2}{\Sigma_\nu(t)}dt, \tag{A.15}$$

subject to (A.8) together with

$$\frac{\gamma}{t_f}\int_0^{t_f}\left(\frac{k_B T(t)}{m} - \Sigma_\nu(t)\right)dt = \mathcal{P},$$

for any given level of power $\mathcal{P}$. Following a variational approach, the Lagrangian for our optimization problem is

$$\mathcal{J} = \frac{T(t)^2}{\Sigma_\nu(t)} + \lambda\left(\Sigma_\nu(t) - \frac{k_B T(t)}{m} + \frac{\mathcal{P}}{\gamma}\right) + \mu\left(\frac{T(t)}{\Sigma_\nu(t)} - \frac{m}{k_B}\right),$$

where λ, μ are the Lagrange multipliers. The stationarity condition for $\Sigma_\nu(t)$ gives

$$\Sigma_\nu(t) = \frac{1}{\sqrt{\lambda}}\sqrt{T(t)^2 + \mu T(t)}.$$

Applying the power constraint we obtain

$$\sqrt{\lambda} = \frac{\int_0^{t_f}\sqrt{T(t)^2 + \mu T(t)}dt}{\frac{k_B}{m}\int_0^{t_f}T(t)dt - \frac{t_f}{\gamma}\mathcal{P}}.$$

In a similar manner as before, the optimal protocol can now be obtained by direct integration of (A.4), to give

$$q(t) = q_0\frac{T(t)^2 + \mu T(t)}{T(0)^2 + \mu T(0)}\exp\left(\frac{2\gamma(t - r(t))}{m}\right), \tag{A.16}$$

where

$$r(t) = \frac{\int_0^{t_f} \sqrt{T(s)^2 + \mu T(s)}ds}{\int_0^{t_f} T(s)ds - \frac{t_f m}{\gamma k_B}\mathcal{P}} \int_0^t \frac{\sqrt{T(s)}}{\sqrt{T(s)+\mu}}ds.$$

It now remains to fix μ by imposing the last constraint, which yields $r(t_f) = t_f$. This is,

$$\int_0^{t_f} \sqrt{T(t)^2 + \mu T(t)}dt \int_0^{t_f} \frac{\sqrt{T(t)}}{\sqrt{T(t)+\mu}}dt = t_f \int_0^{t_f} T(t)dt - \frac{t_f^2 m}{\gamma k_B}\mathcal{P}. \tag{A.17}$$

In general, Eq. (A.17) cannot be solved explicitly for μ. However, it is easily seen that when $\mathcal{P} = 0$, then $\mu = 0$ is a solution. Also, as $\mathcal{P} \to \mathcal{P}^*$, then the solution $\mu \to \infty$. It is worthwhile to point out that the condition $r(t_f) = t_f$ makes the protocol (A.16) continuous for continuous temperature profiles.

For notational convenience, let $\overline{g(t)} = \frac{1}{t_f}\int_0^{t_f} g(t)dt$ denote the average of a periodic function $g(t)$ over the period. Now, putting all of the ingredients together, the optimal $\mathcal{Q}_{\text{qs}}$ is given by

$$\mathcal{Q}^*_{\text{qs}} = \frac{t_f}{\kappa}\left[\frac{\overline{\sqrt{(T(t)+\mu)T(t)}}}{\overline{T(t)} - \kappa\mathcal{P}} \times \overline{\left(\frac{T(t)^{\frac{3}{2}}}{\sqrt{(T(t)+\mu)}}\right)} - \overline{T(t)}\right],$$

where $\kappa = m/(\gamma k_B)$. Consequently, we obtain the following result:

Maximum Efficiency at Fixed Power Under an Arbitrary Temperature Profile

Consider an underdamped particle under a quadratic potential for which the velocity covariance follows (A.4). Over a cycle of period t_f, the maximum efficiency at fixed power $\mathcal{P}$ is given by

$$\eta^*(\mathcal{P}) = \frac{\kappa\mathcal{P}}{\left[\frac{\overline{\sqrt{(T(t)+\mu)T(t)}}}{\overline{T(t)}-\kappa\mathcal{P}} \times \overline{\left(\frac{T(t)^{\frac{3}{2}}}{\sqrt{(T(t)+\mu)}}\right)} - \overline{T(t)}\right]}, \tag{A.18}$$

where $\kappa = m/(\gamma k_B)$ and μ satisfies

$$\int_0^{t_f} \sqrt{T(t)^2 + \mu T(t)}dt \int_0^{t_f} \frac{\sqrt{T(t)}}{\sqrt{T(t)+\mu}}dt = t_f \int_0^{t_f} T(t)dt - \frac{t_f^2 m}{\gamma k_B}\mathcal{P}.$$

(continued)

The optimal protocol is given by

$$q(t) = q_0 \frac{T(t)^2 + \mu T(t)}{T(0)^2 + \mu T(0)} \exp\left(\frac{2\gamma(t - r(t))}{m}\right),$$

where

$$r(t) = \frac{\int_0^{t_f} \sqrt{T(s)^2 + \mu T(s)} ds}{\int_0^{t_f} T(s) ds - \frac{t_f m}{\gamma k_B} \mathcal{P}} \int_0^t \frac{\sqrt{T(s)}}{\sqrt{T(s) + \mu}} ds.$$

Figure A.1 displays the trade-off between power and efficiency for three different temperature profiles. Note that for vanishing power we obtain the maximum possible efficiency, i.e. $\eta^* = 1$, that corresponds to vanishing dissipation. On the other hand, as $\kappa \mathcal{P} \to \kappa \mathcal{P}^* = \mathrm{Var}(\sqrt{T})$ (from (A.12)), the limiting value from (A.18) gives that

$$\eta^* \to \frac{\overline{\sqrt{T(t)}} \mathrm{Var}(\sqrt{T})}{\overline{T(t)^{\frac{3}{2}}} - \overline{T(t)} \times \overline{\sqrt{T(t)}}}. \tag{A.19}$$

This is precisely the efficiency at maximum power obtained by using the optimal protocol (A.11). The limit in (A.19) can also be expressed in terms of the third moment of the square root of the temperature

$$\mu_3(\sqrt{T}) = \frac{1}{t_f} \int_0^{t_f} \left(\sqrt{T(t)} - \overline{\sqrt{T(t)}}\right)^3 dt,$$

specifically,

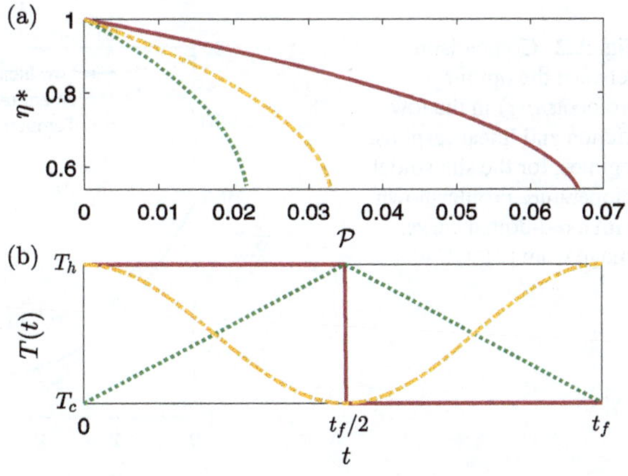

Fig. A.1 (**a**) Maximum efficiency at fixed power for the temperature profiles portrayed in (**b**). (**b**) Temperature profiles with colors and patterns matching the corresponding performance curves in (**a**). Image from [2]

$$\eta^* \to \frac{1}{2 + \frac{\mu_3(\sqrt{T})}{\mathrm{Var}(\sqrt{T})\overline{\sqrt{T(t)}}}}. \tag{A.20}$$

Thus, for any temperature profile such that $\mu_3(\sqrt{T}) = 0$, the limit is $\frac{1}{2}$, which in particular is true for the Carnot-like temperature profile $T(t) = T_{\mathrm{Carnot}}(t)$. We also note that the third moment can be negative, leading to efficiency that exceeds $\frac{1}{2}$, albeit at the cost of decreasing maximum power.

A.3.1 Example: Sinusoidal Temperature Profile

In the following, we compare our results to the ones obtained in the linear regime, where it is assumed that $T_h - T_c \ll T_h + T_c$ [5, 13]. To this end, we consider the sinusoidal temperature profile

$$T(t) = \frac{T_h + T_c}{2} + \frac{T_h - T_c}{2}\cos(\omega t), \tag{A.21}$$

with $\omega \ll \sqrt{\frac{q}{m}}$ so as to ensure that the low friction assumption holds.

We first use our analytical result to recover the expressions obtained in the linear regime. Specifically, assuming $\Delta T = \frac{T_h - T_c}{2}$ is much smaller than $\bar{T} = \frac{T_h + T_c}{2}$, we have that

$$\begin{aligned} q^*(t) &= q_0 + q_0\frac{\Delta T}{\bar{T}}\cos(\omega t) + \mathcal{O}(\Delta T^2), \\ \mathcal{P}^* &= \frac{\gamma k_B}{8m}\frac{\Delta T^2}{\bar{T}} + \mathcal{O}(\Delta T^3), \\ \eta^*(\mathcal{P}^*) &= \frac{1}{2} + \mathcal{O}(\Delta T^2), \end{aligned} \tag{A.22}$$

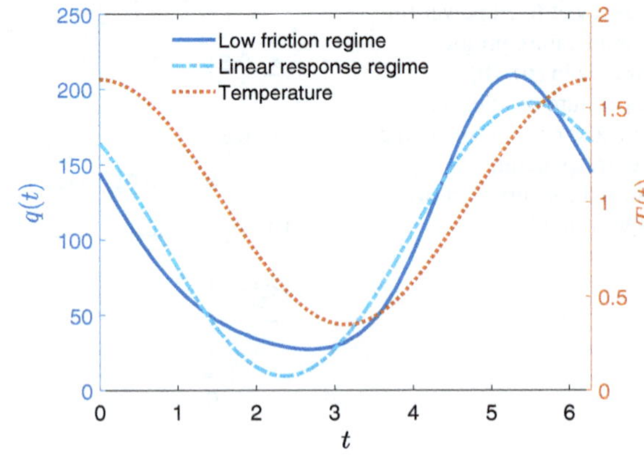

Fig. A.2 Comparison between the optimal protocols $q(t)$ in the low friction and linear response regimes, for the sinusoidal temperature profile shown with a red-dotted curve. Image from [2]

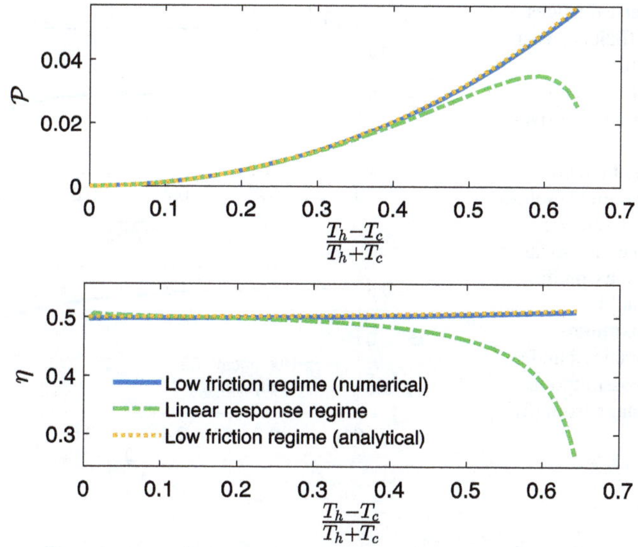

Fig. A.3 Power and efficiency as functions of temperature ratios for two protocols: (i) the optimal protocol obtained under the low friction assumption (A.11) (solid line), and (ii) the optimal protocol obtained in the linear response regime [13, Eqs 29, 30a] (dashed line). For comparison, the analytical expressions for maximum power (A.12) and efficiency (A.19) in the low friction regime are also drawn (dotted curve). Image from [2]

where the first two coincide with the expressions for the optimal protocol and power that are given in [5, 13], in the limit where $\frac{q_0}{m}$ is large enough to satisfy the requirements of the low friction regime. Furthermore, the third one, Eq. (A.22), holds for any temperature profile whose time-dependent component is an odd function of t as long as $\Delta T \ll \bar{T}$.

Moreover, in order to illustrate the utility of our results, we compare the optimality of the protocol obtained here, to the one obtained in the linear regime. Specifically, we solve numerically the equations for the variance (A.1) for two protocols:

(i) the optimal protocol obtained with low friction assumption according to (A.11),
(ii) the optimal protocol obtained in linear regime in [5] and [13, Eqs 29, 30a].

These two protocols, along with the temperature profile, are shown in Fig. A.2. Then, we use the numerical solution to the equations for the variance (A.1) to compute power and efficiency. The result, as the temperature difference $T_h - T_c$ is varied, is shown in Fig. A.3.

In these numerical examples, we made sure $\frac{\gamma}{m} \ll \sqrt{\frac{q}{m}}$ so that the low friction assumption remains valid. As it is expected, the optimal protocol under the linear response regime assumption performs well when the temperature ratio is small,

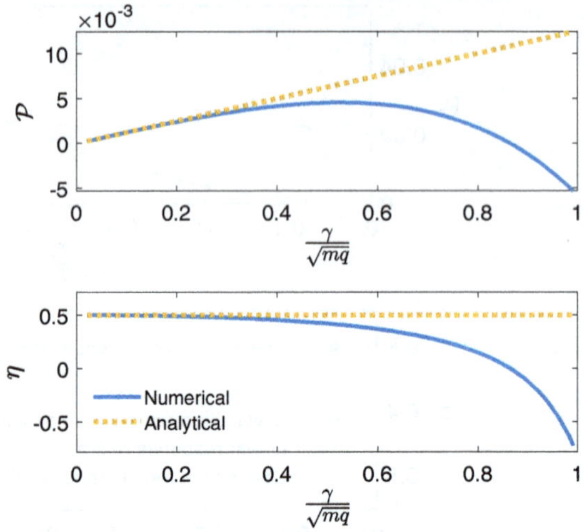

Fig. A.4 Numerical values for power and efficiency as a function of the friction coefficient for the system governed by Eq. (A.1), driven by the optimal protocol (A.11) obtained in the low friction regime (solid line). For comparison, we have included the analytical expressions for maximum power (A.12) and the efficiency at maximum power (A.19) obtained under the low friction assumption (dotted line). Image from [2]

giving a numerical result that agrees with equations (13) and (20). However, as the temperature ratio increases, higher order terms become significant and the protocol is no longer optimal.

We finally highlight how the numerical value of power and efficiency, using the optimal protocol (A.11), deviates from the analytical expressions for maximum power (A.12) and efficiency at maximum power (A.19) with increasing values of the friction coefficient γ. This is shown in Fig. A.4. Specifically, for small values of $\frac{\gamma}{\sqrt{mq}}$, the analytical and numerical results for power and efficiency coincide. For larger values of $\frac{\gamma}{\sqrt{mq}}$ the approximation error becomes significant due to the fact that the condition $\frac{\gamma}{\sqrt{mq}} \ll 1$, and hence, the approximate equipartition relation (A.3), no longer hold.

Appendix B
Optimal Linear Gaussian Information Engine

In Sect. 2.5.1 we learned that having access to information can allow us to extract work from a thermodynamic system. Specifically, even if the free energy of our system is kept constant and we are subject to a unique temperature, the second law still allows us to extract roughly as much work as information we use. In this appendix, we study how to extract the maximum amount of work possible from an overdamped system subject to continuous measurements.

In order to do so, we focus on the case of a quadratic potential function $U(t,x) = \frac{q_0}{2}(x - r_t)^2$, where the location r_t of the center of the potential represents the control input, while the intensity q_0 remains constant. We assume access to continuous measurements of the particle with observation function $h(x) = x$. Thus, the dynamics for the particle and the observation are

$$dX_t = -\frac{q_0}{\gamma}(X_t - r_t)dt + \sqrt{\frac{2k_BT}{\gamma}}dB_t \tag{B.1a}$$

$$dZ_t = X_t dt + \sigma_v dV_t. \tag{B.1b}$$

The objective is to maximize the work output during a cycle of period t_f by designing the control input r_t. We assume the boundary condition $r_0 = r_{t_f} = 0$. We also assume that the initial probability distribution is at equilibrium to disregard any amount of work that can be extracted if the system is not prepared at equilibrium. For the initial potential $U_0(x) = \frac{q_0}{2}x^2$, the equilibrium distribution is Gaussian with zero mean and variance $\Sigma_0 = \frac{k_BT}{q_0}$, denoted by $\mathcal{N}(0, \Sigma_0)$.

In this special linear Gaussian case, the conditional probability distribution of X_t given the observations is Gaussian $\mathcal{N}(m_t, \Sigma(t))$, where the mean and variance evolve according to Kalman-Bucy filter equations [14]

$$dm_t = -\frac{q_0}{\gamma}(m_t - r_t)dt + \frac{\Sigma(t)}{\sigma_v^2}d\xi_t \tag{B.2a}$$

O. Movilla Miangolarra, *Stochastic Thermodynamic Treatment of Thermal Anisotropy*, Springer Theses, https://doi.org/10.1007/978-3-031-68066-3

$$\dot{\Sigma}(t) = -\frac{2q_0}{\gamma}\Sigma(t) + \frac{2k_B T}{\gamma} - \frac{1}{\sigma_v^2}(\Sigma(t))^2, \tag{B.2b}$$

and $d\xi_t = dZ_t - m_t dt$ is the innovation process. In this special case, the work input to the system is

$$W = \int_0^{t_f} q_0(r_t - X_t)\dot{r}_t dt,$$

and the conditional expectation of work given the observations is

$$\mathcal{W}(\mathcal{Z}) = \int_0^{t_f} q_0(r_t - m_t)\dot{r}_t dt,$$

where we replaced X_t with its conditional expectation m_t. Upon integration by parts and utilizing the boundary conditions $r_0 = r_{t_f} = 0$,

$$\mathcal{W}(\mathcal{Z}) = -\frac{q_0^2}{\gamma}\int_0^{t_f} r_t(m_t - r_t)dt + \frac{q_0}{\sigma_v^2}\int_0^{t_f} r_t\Sigma(t)d\xi_t.$$

Finally, taking expectation over the measurement history, the second term disappears and we end up with the following stochastic optimal control problem for maximizing work output

$$\min_u \frac{q_0^2}{\gamma}\mathbb{E}\left[\int_0^{t_f}(u_t^2 - \frac{1}{4}m_t^2)dt\right]$$

$$\text{s.t.} \quad dm_t = -\frac{q_0}{2\gamma}m_t dt + \frac{q_0}{\gamma}u_t dt + \frac{\Sigma_t}{\sigma_v^2}d\xi_t,$$

where we introduced the control input $u_t = r_t - \frac{1}{2}m_t$.

We use the following candidate value function $\mathcal{V}(t, m) = P(t)m^2 + R(t)$ where $P(t)$ and $R(t)$ are time varying parameters to be determined later. We express the objective function as $\frac{q_0^2}{\gamma}\mathbb{E}[J]$ where $J := \int_0^{t_f}(u_t^2 - \frac{1}{4}m_t^2)dt$. Upon adding the zero term $\int_0^{t_f} d\mathcal{V}(t, m_t) - \mathcal{V}(t_f, m_t) + \mathcal{V}(0, m_0) = 0$ to J, and using

$$d\mathcal{V}(t, m_t) = \dot{P}(t)m_t^2 dt + \dot{R}(t)dt - \frac{q_0}{\gamma}P(t)m_t^2 dt + 2\frac{q_0}{\gamma}P(t)m_t u_t dt$$
$$+ P(t)\frac{(\Sigma(t))^2}{\sigma_v^2}dt + 2P(t)m_t\frac{\Sigma(t)}{\sigma_v^2}d\xi_t,$$

we arrive at

$$
\begin{aligned}
J = & \int_0^{t_f} \left[u_t^2 + 2\frac{q_0}{\gamma} P(t) m_t u_t + \left(-\frac{1}{4} + \dot{P}(t) - \frac{q_0}{\gamma} P(t) \right) m_t^2 \right] dt \\
& + \int_0^{t_f} \left(\dot{R}(t) + P(t) \frac{(\Sigma(t))^2}{\sigma_v^2} \right) dt \\
& + \int_0^{t_f} 2P(t) m_t \frac{\Sigma(t)}{\sigma_v^2} d\xi_t - \mathcal{V}(t_f, m_t) + \mathcal{V}(0, m_0).
\end{aligned}
$$

Now we use our freedom to specify $P(t)$ and $R(t)$ to make the first term a complete square and the second term zero:

$$
\begin{aligned}
\dot{P}(t) &= \frac{1}{4} + \frac{q_0}{\gamma} P(t) + \frac{q_0^2}{\gamma^2} P(t)^2, \quad P(t_f) = 0 \\
\dot{R}(t) &= -P(t) \frac{(\Sigma(t))^2}{\sigma_v^2}, \quad R(t_f) = 0.
\end{aligned}
$$

We also set the terminal condition to zero to make $\mathcal{V}(t_f, m_{t_f}) = 0$. The resulting expression for J, after taking the expectation, is

$$
\mathbb{E}[J] = \mathbb{E}\left[\int_0^{t_f} \left(u_t + \frac{q_0}{\gamma} P(t) m_t \right)^2 dt + \mathcal{V}(0, m_0) \right].
$$

The term $\mathcal{V}(0, m_0)$ does not depend on u. Therefore, the optimal control is $u_t = -\frac{q_0}{\gamma} P(t) m_t$, and the optimal value is

$$
\mathcal{W}^* = \frac{q_0^2}{\gamma} \mathbb{E}[J] = \frac{q_0^2}{\gamma} \mathcal{V}(0, m_0) = \frac{q_0^2}{\gamma} R(0),
$$

where we used $m_0 = 0$. The following result follows by noting $R(0) = \int_0^{t_f} P(t) \frac{\Sigma_t^2}{\sigma_v^2} dt$ and changing $P(t) \to \frac{q_0}{\gamma} P(t)$:

Maximum Work Output for the Linear Gaussian Information Engine
Consider a particle governed by the overdamped Langevin equation with quadratic potential and a linear observation model (B.1), and assume that the boundary conditions $r_0 = r_{t_f} = 0$ and the equilibrium initial distribution $\mathcal{N}(0, \frac{k_B T}{q_0})$ hold. The maximum work output over $[0, t_f]$ is

(continued)

$$\mathcal{W}^*_{\text{out}} = -\frac{q_0}{\sigma_v^2}\int_0^{t_f} P(t)(\Sigma(t))^2 dt,$$

and the optimal control is given by $r_t = (\frac{1}{2} - P(t))m_t$, where m_t and $\Sigma(t)$ are the conditional mean and variance of X_t given by the Kalman-Bucy filter equations (B.2) and

$$\frac{\gamma}{q_0}\dot{P}(t) = \left(P(t) + \frac{1}{2}\right)^2, \quad P(t_f) = 0,$$

or, in closed form, $P(t) = \left[\frac{q_0(t_f - t)}{\gamma} + 2\right]^{-1} - \frac{1}{2}$. Moreover, the efficiency at maximum power is

$$\eta = \frac{-2q_0 \int_0^{t_f} P(t)(\Sigma(t))^2 dt}{k_B T \int_0^{t_f} \Sigma(t) dt}.$$

Here we have defined the efficiency of information engines as the ratio between the work output and the amount of information that is available to be used [15], i.e.

$$\eta := \frac{\mathcal{W}_{\text{out}}}{k_B T \mathcal{I}(X_{0:t_f}; Z_{0:t_f})}.$$

In light of (2.24), the efficiency is always smaller than 1. It is also noted that, in order to achieve maximal efficiency, it is necessary that $\mathcal{I}(X_{t_f}; Z_{0:t_f}) = 0$, and thereby, that all available information has been used within the interval.

B.1 Steady-State Analysis

Explicit formulas for average power $\frac{\mathcal{W}^*_{\text{out}}}{t_f}$ and efficiency are obtained in steady-state as $t_f \to \infty$. The steady-state optimal average power is

$$\lim_{t_f \to \infty} \frac{\mathcal{W}^*_{\text{out}}}{t_f} = \frac{q_0 k_B T}{\gamma} \frac{1}{\text{SNR}} \left(\sqrt{1 + \text{SNR}} - 1\right)^2.$$

where $\text{SNR} = \frac{2\gamma k_B T}{q_0^2 \sigma_v^2}$ represents the signal to noise ratio. The limit is obtained using the steady-state values $P_{ss} = -\frac{1}{2}$ and $\Sigma_{ss} = \sigma_v^2(-\frac{q_0}{\gamma} + \sqrt{\frac{q_0^2}{\gamma^2} + \frac{2k_B T}{\gamma \sigma_v^2}})$. In

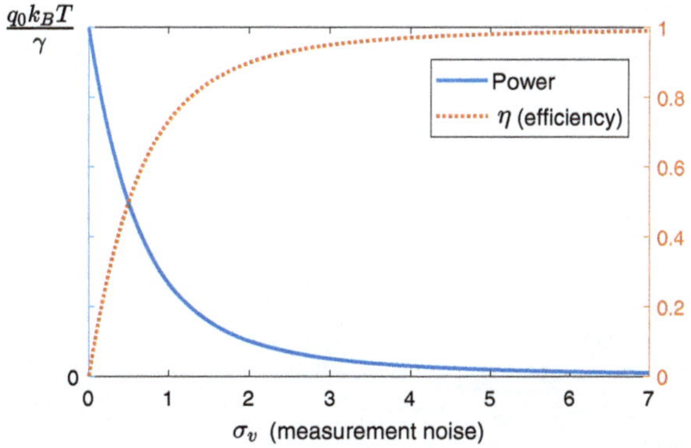

Fig. B.1 Steady-state values for maximum power and efficiency for a linear Gaussian overdamped information machine, as a function of the measurement noise. Image from [16]

particular, as $\sigma_v \to \infty$, power converges to zero (because in this case, effectively, no information is available), and as $\sigma_v \to 0$, power attains its maximum value $\frac{q_0 k_B T}{\gamma}$. The efficiency at steady state becomes

$$\lim_{t_f \to \infty} \eta = \frac{2}{\text{SNR}} \left(\sqrt{1 + \text{SNR}} - 1 \right)$$

Note that as $\sigma_v \to \infty$, the efficiency goes to 1. However, as $\sigma_v \to 0$, the efficiency converges to 0, since the available information is infinite. We illustrate power and efficiency trade-offs as functions of σ_v in Fig. B.1.

Appendix C
Inertialess Gyrating Engine: Proofs

In this section, we provide further technical insights and proofs to Sect. 6.3. We begin by showing that sign-definiteness of the effective torque implies that a system obeying (6.8) has indeed a unique asymptotically stable limit cycle. We continue on by showing that for any θ-periodic torque profile for which a certain continuity condition holds, a finite number of engines always suffices to ensure sign-definiteness of the torque, and thereby stable operation of the system of coupled engines. We then specialize to the case of the Brownian gyrator-based engines with a fixed number of units (three in particular), and we derive alternative sufficient conditions for sign-definiteness of the effective torque as well as explicit expressions for the average angular velocity.

C.1 Sign-Definiteness of Torque Implies a Unique Stable Limit Cycle

Herein we prove that if the torque $\mathcal{T}(\theta)$ is strictly positive for all values of θ, then a unique globally attractive limit cycle exists for any (and hence, for a vanishingly small) amount of inertia I. The basis of the argument to establish existence of such a limit cycle is the Poincaré–Bendixson theorem [17, Page 391, Theorem 2.1], [18, Theorem 9.0.6]. This theorem states that a trajectory of a second-order system, confined in a bounded two-dimensional region of the phase space that contains no fixed points, is either a periodic orbit itself or it converges (asymptotically) to one. The phase space can be a cylinder $[0, 2\pi) \times \mathbb{R}$, as is the case of the system in (6.8).

A fixed point of (6.8) requires that $\omega = 0$ (from the first of the two equations). But then, $\mathcal{T}(\theta) - \Gamma\omega$ cannot vanish, since $\mathcal{T}(\theta) > 0$ for all θ, and hence (6.8) has no fixed points. We observe that any (bounded) region $D = \{(\theta, \omega) \in [0, 2\pi) \times [-M, M]\}$, for sufficiently large M, is positively (in time) invariant. That is, any

O. Movilla Miangolarra, *Stochastic Thermodynamic Treatment of Thermal Anisotropy*, Springer Theses, https://doi.org/10.1007/978-3-031-68066-3

trajectory that begins in D is confined within D for all times. Thus, by the Poincaré–Bendixson theorem, there exists an asymptotically attractive periodic orbit.

We now argue that the claimed periodic orbit is in fact unique, i.e., it represents a globally attractive stable limit cycle. We follow a similar argument to that of Sect. 6.2.1. Starting from a point $(\theta = 0, \omega(0))$ that lies on a period orbit, we integrate $I\dot{\omega} = \mathcal{T}(\theta) - \Gamma\omega$ over the cycle $\theta \in [0, 2\pi)$. The integral of the left-hand side is

$$\int_0^{2\pi} I\dot{\omega}d\theta = \int_0^{t_c} I\dot{\omega}\omega dt = \frac{I}{2}\left(\omega(t_c)^2 - \omega(0)^2\right) = 0,$$

where t_c is the time-duration of a cycle. Integrating the right-hand side now gives that

$$\int_0^{2\pi} \omega d\theta = \frac{2\pi \langle \mathcal{T} \rangle}{\Gamma}, \tag{C.1}$$

where $\langle \mathcal{T} \rangle$ is a (fixed) constant that only depends on the shape of the torque profile. Since trajectories do not cross, Eq. (C.1) can only be satisfied by a unique periodic orbit.

C.2 Number of Gyrating Engines Required to Dispense with Inertia

Let us consider gyrating engines obeying (6.8). Following two different approaches we show that provided the torque profile $\mathcal{T}(\theta)$ satisfies $|\mathcal{T}(\theta + \Delta) - \mathcal{T}(\theta)| < L|\Delta|$ for all θ, Δ and with $L < \infty$ (i.e., it is Lipshitz) and provided the average torque over a cycle is not zero (and which, without loss of generality, is assumed positive), there is an integer m so that m equidistantly-coupled engines ensure a globally attractive limit cycle. In other words, we establish that under natural and mild conditions on the torque profile, a finite number of coupled Brownian gyrating engines is always sufficient to maintain a stable limit cycle for *any* set of parameters.

To establish the claim we show that a finite number of coupled engines is sufficient to ensure strictly positive torque for all values of the angular position θ. Assuming that $\mathcal{T}(\theta)$ is Lipschitz and periodic, we consider the Fourier series expansion

$$\mathcal{T}(\theta) = c_0 + \sum_{k=1}^{\infty} a_k \cos(k\theta) + b_k \sin(k\theta). \tag{C.2}$$

For m equidistantly-coupled engines the effective torque is

$$\begin{aligned}\mathcal{T}_m(\theta) &= \frac{1}{m}\sum_{\ell=0}^{m-1}\mathcal{T}\left(\theta - \tfrac{2\pi}{m}\ell\right)\\ &= c_0 + \frac{1}{m}\sum_{k=1}^{\infty}\sum_{\ell=0}^{m-1} a_k\cos(k(\theta - \tfrac{2\pi}{m}\ell)) + b_k\sin(k(\theta - \tfrac{2\pi}{m}\ell))\\ &= c_0 + \sum_{k=1}^{\infty} a_{km}\cos(km\theta) + b_{km}\sin(km\theta) = \sum_{k=-\infty}^{\infty} c_{km}e^{ikm\theta},\end{aligned}$$

where $c_k = \frac{1}{2}(a_k + ib_k)$ for $k > 0$ and $c_k = \overline{c_{-k}}$ for $k < 0$. The third equality follows from cancellation, due to phase difference, of all terms with indices that are not multiples of m. Since $\mathcal{T}(\theta)$ is Lipschitz, the amplitude of the harmonics decays faster than k^{-1} and the series $\{|c_k| : k > 0\}$ is summable, see e.g., [19]. Thus, there exists an m such that $c_0 > \sum_{k\neq 0}|c_{km}|$, and for this m, $\mathcal{T}_m(\theta) > 0$ for all θ.

An alternative argument can be drawn as follows. Denote by L the torque's Lipschitz constant, i.e., $L = \inf\{\kappa \mid |\mathcal{T}(\theta + \Delta) - \mathcal{T}(\theta)| < \kappa|\Delta|$, for all $\theta, \Delta \in [0, 2\pi]\}$. Then, $\mathcal{T}_m(\theta)$ is also Lipschitz with Lipshitz constant $\leq L$. It is also periodic with period $2\pi/m$ and average c_0, which we assume positive. Let ϑ_0 be such that $\mathcal{T}_m(\vartheta_0) = c_0$, which always exists since $\mathcal{T}_m$ is continuous. Then, over a period $\theta \in \left[\vartheta_0 - \frac{\pi}{m}, \vartheta_0 + \frac{\pi}{m}\right]$, $\mathcal{T}_m(\theta)$ takes values in the interval $\left[c_0 - \frac{L\pi}{m}, c_0 + \frac{L\pi}{m}\right]$. Thus, if we take $m = \lceil\frac{L\pi}{c_0}\rceil$, that is, we take the smallest integer m such that $m \geq \frac{L\pi}{c_0}$, it follows that $\mathcal{T}_m(\theta) > 0$ over the period, and hence for all θ. Note that the number $m = \lceil\frac{L\pi}{c_0}\rceil$ of the needed engines is tight when $\mathcal{T}(\theta)$ has the shape of a triangular wave with slope L and period 2π.

C.2.1 Alternative Analysis for the Brownian Case

Finally, we derive a condition for three coupled Brownian gyrating engines ($m = 3$) to suffice for sustained limit cycle operation. We do so by expanding the dimensionless torque for the Brownian gyrating engine in the dimensionless parameter β. This parameter controls the variation of the capacitance and, by expanding around zero, we assume that this variation is small. That is, we assume that our system is within the linear response regime. The expansion gives

$$\frac{\mathcal{T}^B(\theta)}{k_B\bar{T}} = f_1(\theta)\beta + f_2(\theta)\beta^2 + \mathcal{O}(\beta^3),$$

where $f_1(\theta)$ depends on θ through terms linear in $\sin(\theta)$ and $\cos(\theta)$, while $f_2(\theta)$ contains second harmonics and a constant term. Terms independent of β vanish, since for $\beta = 0$ energy cannot be extracted from the system. Therefore, up to second-order in β, the only term that contributes to the average torque is $f_2(\theta)\beta^2$, whose average value over a cycle can be computed to be

$$\frac{1}{k_B\bar{T}}\int_0^{2\pi}\mathcal{T}^B(\theta)d\theta \approx \frac{\beta^2}{2\pi}\int_0^{2\pi} f_2(\theta)d\theta = \frac{\sqrt{3}\Delta T}{64\bar{T}}\beta^2.$$

Consequently, if two engines are coupled, the first order term in β vanishes, whereas, if three engines are coupled the remaining terms are of third order or higher. Thus, provided

$$\frac{\sqrt{3}\Delta T}{64\bar{T}} \gg \beta,$$

the constant term dominates over higher-order terms and a globally attractive limit cycle operation is present for the three coupled engines. In that case, the average angular velocity can be approximated by

$$\langle\omega\rangle \approx \frac{\sqrt{3}k_B\Delta T}{64\Gamma}\beta^2.$$

References

1. Yanagida, T., Ueda, M., Murata, T., Esaki, S., Ishii, Y.: Brownian motion, fluctuation and life. Biosystems **88**(3), 228–242 (2007)
2. Movilla Miangolarra, O., Fu, R., Taghvaei, A., Chen, Y., Georgiou, T.T.: Underdamped stochastic thermodynamic engines in contact with a heat bath with arbitrary temperature profile. Phys. Rev. E **103**, 062103 (2021)
3. Holubec, V., Steffenoni, S., Falasco, G., Kroy, K.: Active brownian heat engines. Phys. Rev. Res. **2**(4), 043262 (2020)
4. Blickle, V., Bechinger, C.: Realization of a micrometre-sized stochastic heat engine. Nat. Phys. **8**(2), 143–146 (2012)
5. Bauer, M., Brandner, K., Seifert, U.: Optimal performance of periodically driven, stochastic heat engines under limited control. Phys. Rev. E **93**(4), 042112 (2016)
6. Bauer, M., Brandner, K., Seifert, U.: Thermodynamics of micro- and nano-systems driven by periodic temperature variations. Phys. Rev. X **5** (2015)
7. Brandner, K., Saito, K.: Thermodynamic geometry of microscopic heat engines. Phys. Rev. Lett. **124** (2020)
8. Dechant, A., Kiesel, N., Lutz, E.: Underdamped stochastic heat engine at maximum efficiency. Europhys. Lett. **119**(5) (2017)
9. Zöller, N.: Optimization of Stochastic Heat Engines in the Underdamped Limit. Springer, Berlin (2017)
10. Dechant, A., Kiesel, N., Lutz, E.: All-optical nanomechanical heat engine. Phys. Rev. Lett. **114** (2015)
11. Gomez-Marin, A., Schmiedl, T., Seifert, U.: Optimal protocols for minimal work processes in underdamped stochastic thermodynamics. J. Chem. Phys. **129** (2008)
12. Schmiedl, T., Seifert, U.: Efficiency at maximum power: an analytically solvable model for stochastic heat engines. Europhys. Lett. **81**(2) (2007)

13. Fu, R., Movilla Miangolarra, O., Taghvaei, A., Chen, Y., Georgiou, T.T.: Harvesting energy from a periodic heat bath. In: 2020 59th IEEE Conference on Decision and Control (CDC) (2020)
14. Kalman, R.E., Bucy, R.S.: New results in linear filtering and prediction theory. J. Basic Eng. **83**(1), 95–108 (1961)
15. Bauer, M., Abreu, D., Seifert, U.: Efficiency of a Brownian information machine. J. Phys. A Math. Theor. **45**(16), 162001 (2012)
16. Taghvaei, A., Movilla Miangolarra, O., Fu, R., Chen, Y., Georgiou, T.T.: On the relation between information and power in stochastic thermodynamic engines. IEEE Control Syst. Lett. **6**, 434–439 (2022)
17. Coddington, E.A., Levinson, N.: Theory of Ordinary Differential Equations. Tata McGraw-Hill Education, New York (1955)
18. Wiggins, S.: Introduction to Applied Nonlinear Dynamical Systems and Chaos, vol. 2. Springer, Berlin (2003)
19. Zygmund, A.: Trigonometric Series, vol. 1. Cambridge University Press, Cambridge (2002)

Milton Keynes UK
Ingram Content Group UK Ltd.
UKHW051246031124
450475UK00001B/3

9 783031 680656